DINOSAURS IN AUSTRALIA

Reconstruction of the Australian Early Cretaceous ichthyosaur *Platypterygius australis*. This dolphin-like predator reached lengths of over 5 m.

DINOSAURS IN AUSTRALIA

Mesozoic Life from the Southern Continent

Benjamin P. Kear, Uppsala University and
Robert J. Hamilton-Bruce, South Australian Museum

Artistic reconstructions by Josh Lee

CSIRO
PUBLISHING

National Library of Australia Cataloguing-in-Publication entry

Kear, Benjamin P.

Dinosaurs in Australia: Mesozoic life from the Southern Continent/by Benjamin P. Kear and Robert J. Hamilton-Bruce.

9780643100459 (pbk.)
9780643101692 (pdf)
9780643102316 (epub)

Includes bibliographical references and index.

Dinosaurs – Australia.
Fossils – Australia.
Paleontology – Australia – Mesozoic.

Hamilton-Bruce, Robert J.

567.90994

Published by

CSIRO PUBLISHING
36 Gardiner Road, Clayton VIC 3168
Private Bag 10, Clayton South VIC 3169
Australia

Telephone: [+613] 9545 8555
Local call: 1300 788 000 (Australia only)
Fax: +61 3 9662 7555
Email: csiropublishing@csiro.au
Website: www.publishing.csiro.au

Front cover: A group of dromaeosaur-like theropods scavenge turtle and fish carcasses in a dried riverbed near what is now Lightning Ridge in northern New South Wales. (Josh Lee)

Back cover: Opalised turtle and other reptile remains from the Lower Cretaceous Griman Creek Formation of Lightning Ridge. (Photographer: Robert Smith)

Set in 10/13 ITC New Baskerville and Adobe Myriad Pro
Edited by Adrienne de Kretser, Righting Writing
Cover and text design by James Kelly
Typeset by Desktop Concepts Pty Ltd, Melbourne
Printed by Ingram Lightning Source

Mar26_RP_ILS

FOREWORD

The ancient world of the dinosaurs, technically known as the Mesozoic, has an almost magical ability to capture the imagination. Yet despite its iconic popular imagery, much of what we know about this era of Earth's history is based on tantalisingly rare and often fragmentary fossilised remains. This is especially true of Australia's Mesozoic record, which has proven particularly difficult to reconstruct because low topographic relief and severe weathering have limited the availability of relevant fossils. In addition, only a small number of palaeontologists have ever been able to actively explore the remote rock exposures of this vast and sparsely populated continent. However, a flurry of research in the past few decades has at last brought the uniqueness and evolutionary significance of the Australian Mesozoic biota to international attention. Some of the most important discoveries include the diverse Cretaceous polar assemblages from the south coast of Victoria, which have shown that ancient Australia was both an evolutionary cradle for modern lineages – possibly including our own group, the placental mammals – and a final refuge for organisms whose relatives had long since gone extinct elsewhere. The spectacular opalised fossils from the dry red heart of northern South Australia have revealed a time when much of the present landmass was covered by a freezing inland sea and nourished one of the richest Cretaceous marine ecosystems yet known from the southern hemisphere.

Most of Australia's Mesozoic rock exposures date from the early part of the Cretaceous period, a stratigraphical interval poorly represented on other continents. Not surprisingly, therefore, the majority of reports on Australian Mesozoic organisms relate to finds from these sediments. However, intensive exploration of formerly inaccessible areas has resulted in the recovery of even older animal and plant fossils that have pushed the record back to the dawn of the Mesozoic. It has also demonstrated that then, as now, Australian fauna and flora have been extensively shaped by geographic and climatic isolation.

We are only just beginning to understand the long and varied history of Mesozoic life from Australia; through this volume, the authors have contributed one of the most important recent advances. This comprehensive compendium of the entire Australian Mesozoic record incorporates the latest scientific information on faunas, their associated floras, and fossil localities, and thus represents an invaluable primary resource for both specialists and non-technical readers. I recommend it to you as a doorway into an incredible period in the Earth's geological past, and a glimpse of one of the most fascinating chapters in the chronicle of life on the Australian continent.

Tim Flannery
December 2009

Reconstructed head and neck of a male (top) and female (bottom) Early Cretaceous plesiosaur, *Umoonasaurus demoscyllus*. The thin crests along the snout and above the eyes might have been used for display and were probably brightly coloured. Image after Kear *et al*. (2006).

ACKNOWLEDGMENTS

This book would not have been possible without the continuous support of Marilyn Kear, Jonas Kear, Philip and Judy Kear, and Alexandra Gaut – thank you all for being there.

Deepest thanks to the staff at CSIRO Publishing for enabling this project to move forward and for seeing it through to fruition with much patience and professionalism.

Alexandra Gaut, Anne Hamilton-Bruce, Marilyn Kear, Neville Pledge, Tom Rich and several anonymous reviewers read and greatly improved drafts of the text.

We are extremely grateful to those who generously contributed images: Neville Alley, Jo Bain, Jenni Brammall, Andrew Cody, David and Judy Elliott, Jim Gehling, Henk Godthelp, Scott Hocknull, John Long, Robert Smith, Tom Rich, Alex Ritchie and Anne Warren. Several institutions and businesses also provided access for photography: the Australian Age of Dinosaurs Museum, the Australian Museum, the Australian Opal Centre, Cody Opals, Museum Victoria, the National Opal Collection, the Natural History Museum, the Northern Territory Museum and Art Gallery, Primary Indistries South Australia (PIRSA), the Queensland Museum, the Royal Adelaide Hospital, the South Australian Museum, Umoona Opal Mine and Museum, and the Western Australian Museum.

Many other individuals gave freely of their time, knowledge and assistance with information and access to fossil specimens: Yanni Athanasiadis, Andrew Cody, Gavin Dally, Tim Flannery, Scott Hocknull, Tom and Sharon Hurley, Robert Jones, Mike Lee, David Pickering, Dale Price and Tanja Burk, Natalie Schroeder, Mikael Siversson, Elizabeth Smith, Dick Suter and John Suter, and Yong Ye Zhen.

CONTENTS

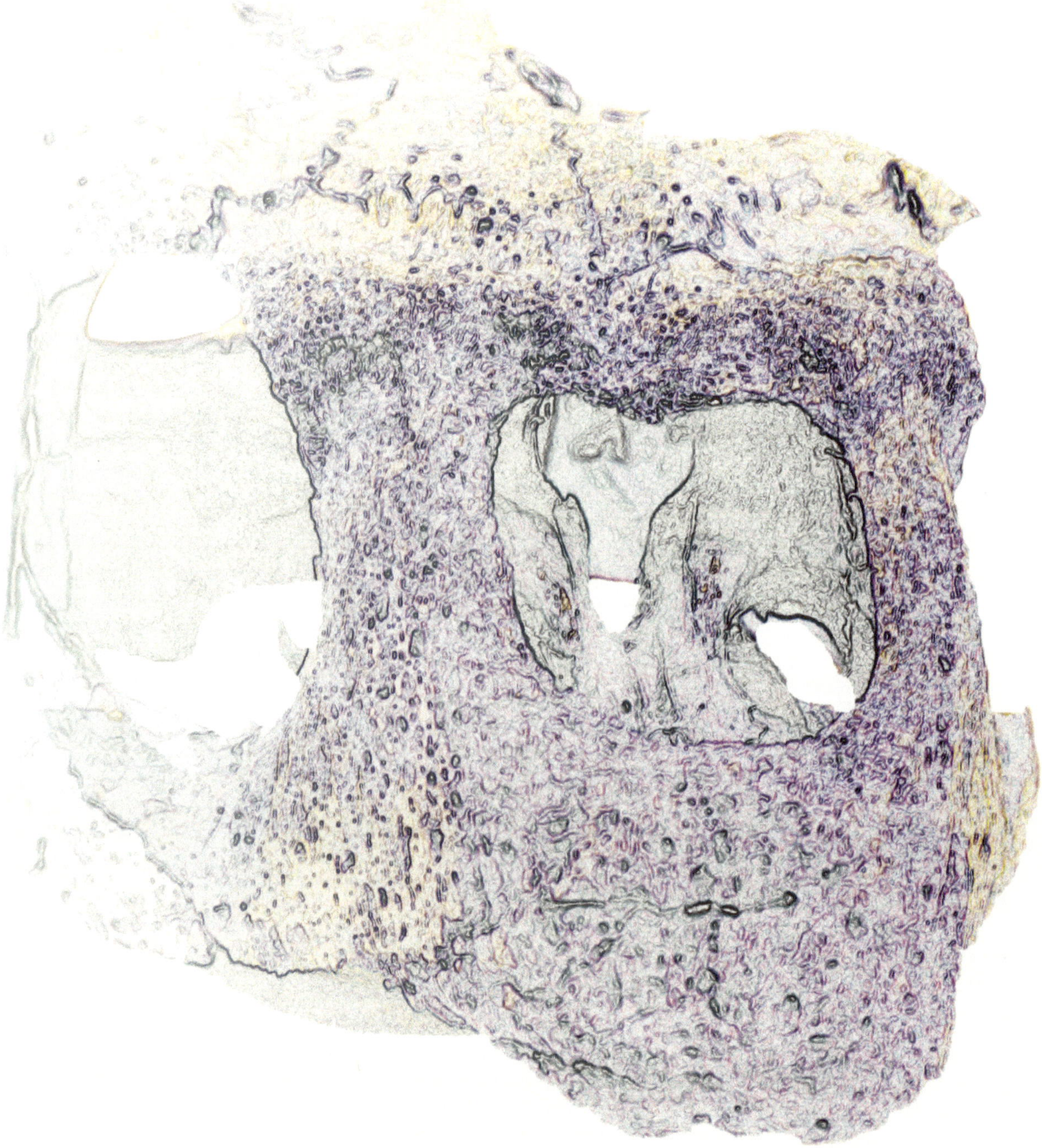

1

INTRODUCTION

Palaeontology – the study of fossils – is arguably one of the most readily accessible sciences. Its re-creation of lost worlds populated by real-life monsters provides fuel for the imaginations of adult and child alike. The painstaking research that lies behind this popular imagery is no less fascinating. The basic work of palaeontologists involves the meticulous excavation, preparation and interpretation of fossils. Fossils are the remains of animals and plants found embedded in rocks either as petrified hard parts or as moulds, casts or tracks. In many cases, fossils are extremely common and can be found by anybody who knows where to look. This book aims to encourage this interest by providing a comprehensive overview of some of the most spectacular Australian fossils – those from the Mesozoic Era, the 'Age of Dinosaurs'. The Mesozoic lasted from ~251 to 65 million years ago, a time when dinosaurs roamed the land, giant sea monsters cruised the oceans and the earliest ancestors of Australia's unique modern animals and plants began to appear. The ancient geography and environments of Australia during the Mesozoic are difficult to imagine. For example, the current landmass did not exist as an island continent, but formed an isolated south-easterly peninsula of a vast supercontinent called Gondwana. This comprised present-day South America, Africa and Madagascar, India, Antarctica and Australia, with the Australian portion situated close to the southern polar circle and, at times, experiencing temperatures close to freezing. Indeed, evidence exists for seasonal sea ice and even permafrost. Such conditions are virtually unique to the Australian Mesozoic record and thus constitute one of the most unusual and globally significant habitats from the Age of Dinosaurs. During much of the later Mesozoic, Australia was nearly totally submerged beneath a shallow inland sea. The muddy sediments from this ancient ocean form an extremely plentiful source of fossils, and are one of the most complete records of life from this period anywhere in the world.

Much of the research on Australian Mesozoic fossils is in the form of technical papers, university theses and conference abstracts that can be difficult to access and collate. This book aims to encapsulate this literature by summarising current work on Australian Mesozoic vertebrates (backboned animals), while also providing a palaeobiogeographical overview of the associated invertebrates (animals without backbones) and different land plant settings. We hope that this information, together with the graphic interpretations and reconstructions, makes this book a useful primary reference, and spurs people to further investigate Australia's amazing Age of Dinosaurs.

How this book is arranged

Understanding the context of fossils and their ancient environments requires some basic background information. This is provided in the introductory chapter, which examines the nature of sedimentary rocks and how fossils are formed, the concept of geological time and the dating of fossil-bearing deposits, the mobile nature of the Earth's crust and the theory of plate tectonics, how fossils are recovered, prepared and interpreted, and lastly how we understand the evolutionary relationships of organisms through time.

Chapter 2 describes Australian floras, faunas and environments during the Mesozoic and includes an overview of the relevant time-scales and maps showing the changing geographic placement of the continent.

The following chapters present a more technical overview of Australian Mesozoic fossils. Each of the major time periods – the Triassic, Jurassic and Cretaceous – is examined in detail, with a brief introduction to their respective geological context, environments and important rock units/localities. The Triassic is discussed in Chapter 3, which examines life from the beginning of the Age of Dinosaurs. The major types of land plants, invertebrates and vertebrates are reviewed and special emphasis is placed on the vertebrate fossil record as a distinctive indicator of time-frames. This format is also applied to the Jurassic in Chapter 4, and the Cretaceous in Chapters 5 to 7. Because the vast majority of Australian Mesozoic fossils are of Cretaceous age, this period is divided into three parts. Chapter 5 examines the Early Cretaceous marine record, representing the time when a vast inland sea inundated much of the Australian continent. Chapter 6 looks at fossils from the contemporary Lower Cretaceous terrestrial and freshwater deposits. Finally, Chapter 7 investigates the remains of life from the end of the Age of Dinosaurs, after the inland sea had disappeared.

Wherever possible, graphic interpretations are included to assist with understanding the relevant anatomical structures. Reconstructions, based directly on the fossil information, show how the organisms might have looked in life. This book frequently uses technical terminology. Therefore a glossary of important terms is included at the end, together with a list of relevant scientific articles to guide further specialist reading.

Sedimentary rocks and the formation of fossils

Ever since the rocks of the Earth's crust first solidified from a molten state some 4.6 billion years ago, the forces of erosion have been wearing them away, reducing them to sediments. These sediments are carried and deposited by rivers, lakes and seas into hollows in the Earth's crust where they eventually compact to form successive layers of sedimentary rock. These may be uplifted into mountain ranges or downwarped into basins and, through the actions of wind and water, provide new areas upon which the cyclical forces of erosion and deposition can work. Indeed, the Earth's surface as we see it now is a product of these continuous cycles with rock formation, erosion and crustal movements all contributing to the development of our modern geography.

The successive layers of sedimentary rock can be read much like the pages of a book. By examining the chronicle of their formation, we can understand the forces shaping the surface of the Earth. Similarly, to appreciate the large-scale processes affecting the development of life we must first understand its history. The best way to do that is with fossils, the remains of animals and plants preserved in sedimentary rocks. As sediment layers are laid down in rivers, lakes and seas, parts of animals and plants can become incorporated and thus be protected by rapid burial, low oxygen conditions or acidic/mineral-rich water, from decay. As the sediments are compacted and turned into rock, the organic remains buried within them become mineralised and are preserved in a petrified state.

The most common types of fossils are microfossils (microscopic fossils). These typically include the tiny calcareous or silicified skeletons of marine plankton or the spores and pollen grains produced by plants. Microfossils are particularly important for determining the age of sedimentary rock strata. This is because their small size allows for easy subsurface sampling using drillcores. In the case of spores and pollen, their resistant waxy outer covering ensures preservation under a variety of conditions.

This book is primarily concerned with macrofossils (those visible with the naked eye), which include remains such as shells, wood or bones. These can be preserved in a number of ways depending on their nature and the conditions that prevailed during and after their entombment in sediment. One example frequently encountered in marine strata is preservation in rounded limestone boulders called concretions. These are produced when decomposing organic material prompts a chemical change in the surrounding sediment, forcing calcium carbonate

There are many kinds of planktonic fossils. Common examples include dinoflagellates and foraminiferans. These minute single-celled eukaryotic (complex-celled) organisms move using whip-like threads or extrudible pseudopodia. The fossil dinoflagellate skeletons shown under magnification were extracted from Lower Cretaceous high-latitude marine sediments in northern South Australia. Dinoflagellates have been used extensively, together with fossilised plant pollen and spores, to determine the relative ages of Mesozoic strata in Australia. Images courtesy Neville Alley and PIRSA.

($CaCO_3$) out of solution. This then precipitates as limestone, which encases the hard organic remains, sometimes with phosphatised soft tissues and other inclusions such as gut contents.

Fine internal structures can also be fossilised by replacement with minerals such as silica (SiO_2). This is the primary component in petrified wood and opalised fossils. Opal fossils are very commonly found in Australia. They occur when hydrated silica ($SiO_2.nH_2O$) directly replaces the organic material, thus preserving a natural replica of the fine surface detail, and sometimes preserving complex internal structures. The process of opal formation is poorly understood, although it may be linked to weathering and cyclical fluctuations of regional watertables (prompting dissolution of opal from silica-rich groundwater), or the actions of bacteria living in the sediment (decomposing organic material and secreting silica as a waste product).

Detailed external preservation can also be facilitated by fine mud or volcanic ash. When this settles in still water bodies such as lakes or poorly oxygenated deep oceans it can act as an inhibitor of decay, creating a high-resolution impression of the animal or plant. This form of fossilisation occasionally preserves remnants of the soft tissues as a thin carbonaceous layer. Common examples of carbonised impressions include delicate leaves and fern fronds, which may show evidence of microsopic surface structures replicated in the mud long after the actual plant material has decayed. Original fossil plant matter is also preserved as coal. This is the carbonaceous remains of plants that grew in localised swampy environments over long periods of time. The immense volumes of accumulating plant matter compact under their own weight to form a coal seam. None of the original plant structure is left within the coal (having been destroyed by compaction); nevertheless, interspersed layers of sediment (usually shale or

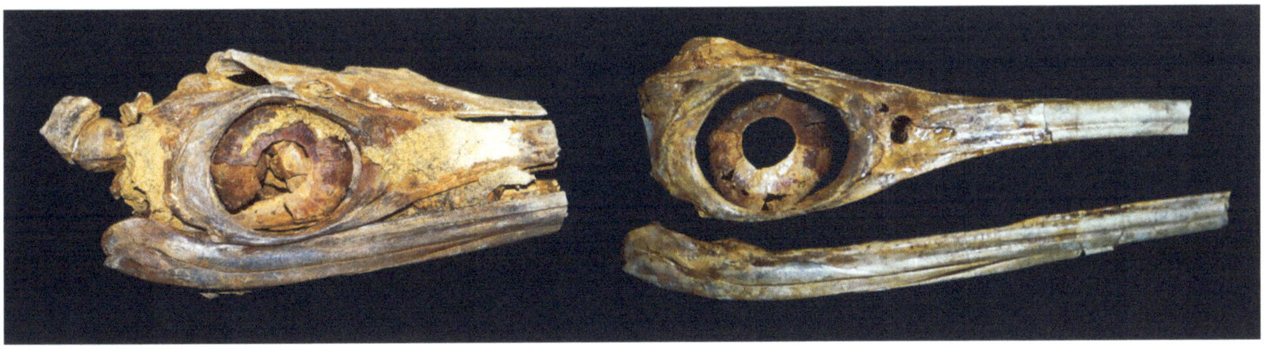

Vertebrate fossils preserved in limestone concretions can be treated with acetic acid to reveal the encased bones. This ichthyosaur skull is shown both during the acid leaching process (left) and after complete extraction (right).

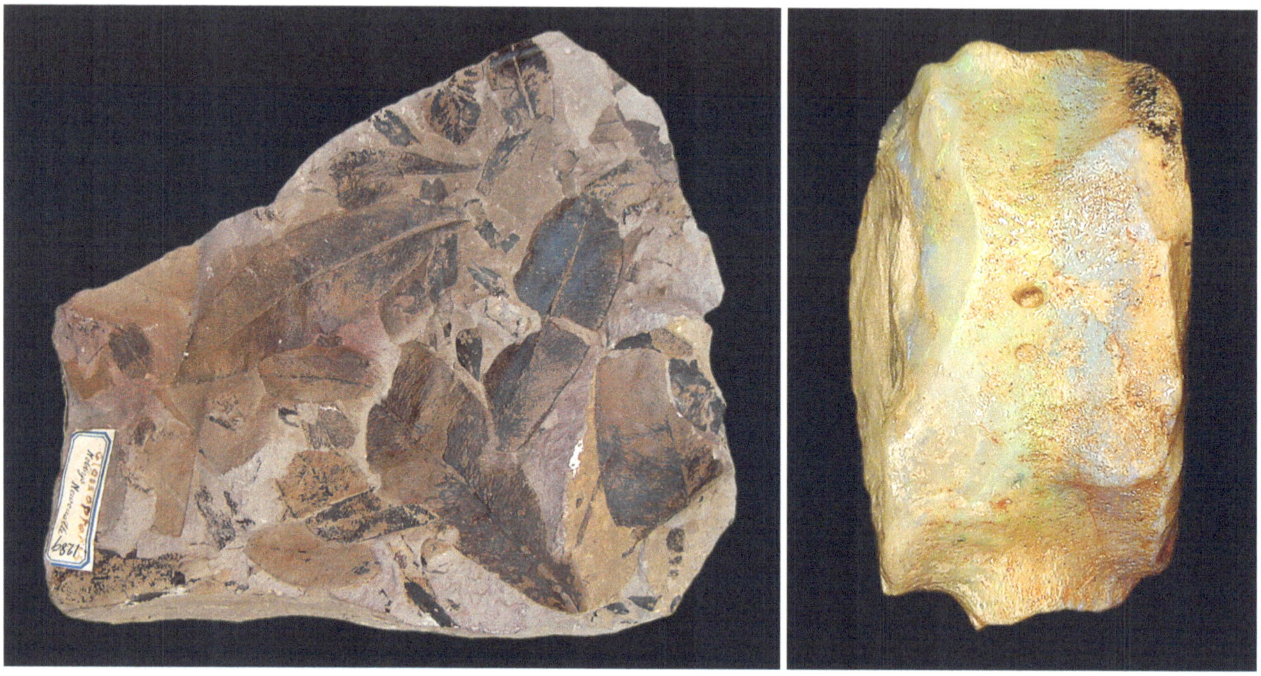

Types of fossil preservation. (Left) Carbonised leaves belonging to the Permian gymnosperm *Glossopteris* from Newcastle in New South Wales, Australia. (Right) An opalised plesiosaur vertebra from Andamooka, South Australia.

fine sandstone) can preserve fossils of animals and plants living in the original swampy habitats.

A preservation type that does not leave any original organic material is moulding and casting.

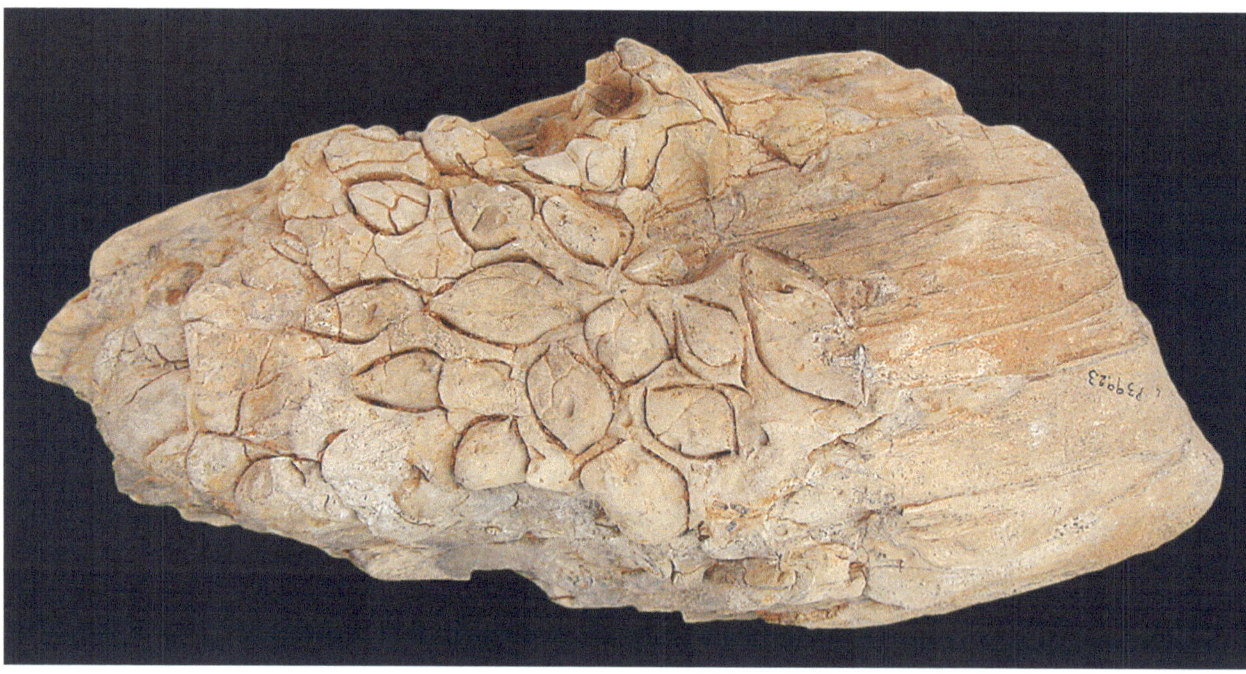

Shell moulds of the mussel-like bivalve *Eyrena* encrusting a piece of fossilised driftwood. This specimen comes from the Lower Cretaceous marine rocks of South Australia.

Mould and cast fossils form when sediment fills a cavity left by the original organic material. (Left) Impressions of the mussel-like bivalve *Eyrena*. Weathering has removed the original shell (reconstructed) to leave an impression in the limestone. (Right) An impression of an ammonite shell that has been naturally cast in limestone. Part of the cast has been destroyed by fracturing and erosion on the surface.

This occurs when organic remnants such as shells are destroyed via dissolution in acidic groundwater or through the effects of erosion, but leave a mould of their external surface in the surrounding rock. Frequently, mud or minerals may fill parts of the structure (e.g. the shell), leaving a cast of the internal surfaces.

A similar moulding and casting process is involved in the preservation of plant and animal traces such as root casts, burrows, feeding marks and

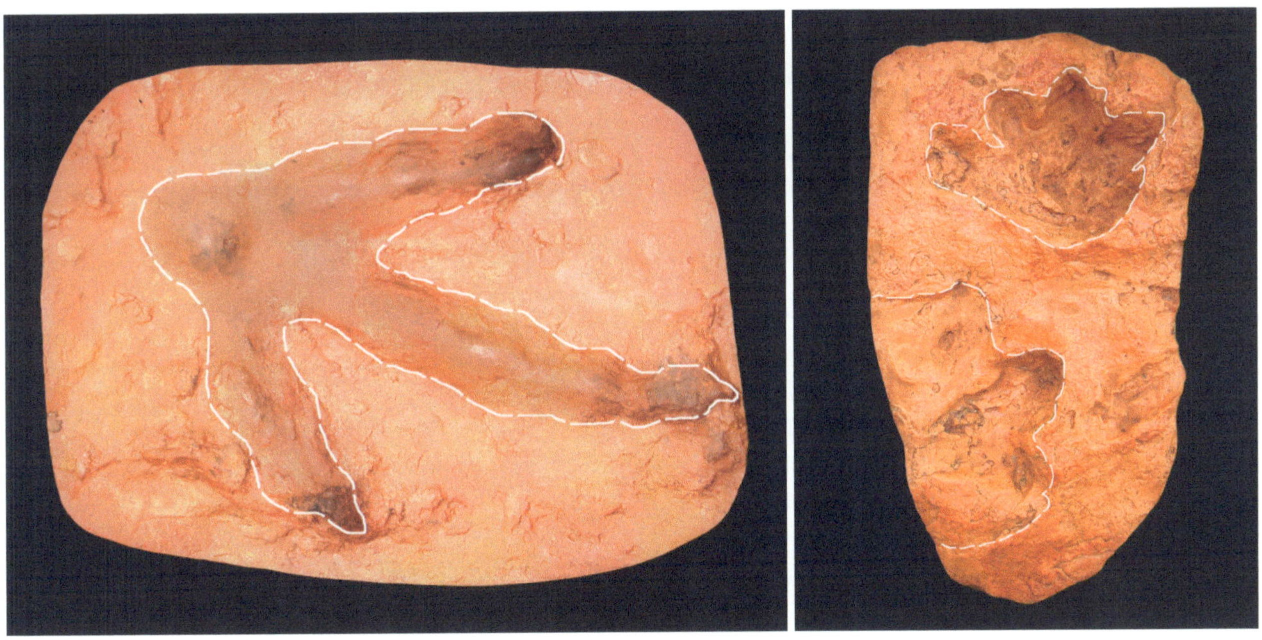

Trace fossils such as footprints are an important source of information about locomotion and behaviour in extinct organisms. (Left) A large theropod footprint from the Lower Cretaceous Broome Sandstone near Broome in far north-western Western Australia. (Right) Possible stegosaur footprints, also from the Broome Sandstone of Western Australia.

footprints. Collectively these are termed 'trace fossils' and they can be useful in determining aspects of behaviour in extinct animals. Trace fossils are generally preserved when sediments are deposited quickly over moisture-rich substrates. For example, when an animal walks across a soft muddy surface the weight of its body creates a series of depressed hollows or tracks showing the impact of its feet. The subsurface sediments are also deformed by the downward pressure of the foot, producing a series of underprints or 'ghost tracks' in the underlying sediments. As more mud is deposited over the exposed trackway (perhaps by floodwaters), the original footprints act as a mould from which a natural cast is made in the overlying layer. As the mud is compacted into rock and exposed at the surface by erosion, the tracks may again become visible as positive or negative reliefs on the successive rock strata. Less-defined moulds and casts can also be left by the underprints.

Geological time and the fossil record

The ages of the Earth, from its initial formation 4.6 billion years ago to today, constitute the framework known by geologists as 'geological time'. The invention of this concept in the 19th century was prompted by a growing understanding of fossils and it led to the formulation of a geological time-scale, which divides the ages of the Earth into a series of geochronological units called (from longest to shortest) eons, eras, periods, epochs, ages and chrons. These refer purely to segments of time and are often further broken down into 'early' or 'late' subdivisions (e.g. Early Triassic). Another set of equivalent units was developed to more accurately correlate time divisions with rock strata; these are subdivided into 'Lower' or 'Upper' in reference to the successive rock layers (e.g. Lower Triassic). Such chronostratigraphical blocks of time are defined by fossils and use the terms 'eonotherm', to equate to an eon, 'eratherm' to an era, 'system' to a period, 'series' to an epoch, 'stage' to an age and 'chronozone' or 'zone' to a chron.

The names given to each segment of the geological time-scale are based on various sources, including the places where important rock types are found or the characteristics of those rocks. For example, the first part of the Age of Dinosaurs, the Triassic period, means 'three layers' and refers to the three main Triassic rock subdivisions found in Europe. The succeeding Jurassic is named after the Jura mountains in Europe where fossils of this age were first recognised. The Cretaceous is named after the Latin word 'creta '(chalk), and refers to the Cretaceous chalk cliffs along the south coast of England.

Each of the time units from the geological time-scale is based on a particular suite of diagnostic fossils. Consequently, if comparable fossils are found in rocks elsewhere, they can be used to provide a correlated age for geographically widely separated strata. Therefore fossils are a vital resource for dating sedimentary rocks and, when coupled with other absolute methods of rock dating (see below),

Stromatolites are some of the earliest known reef-building organisms. The examples shown here are (left) the Cambrian stromatolite *Trezona* from the Flinders Ranges of South Australia, and (right) its cross-section. Images courtesy James Gehling.

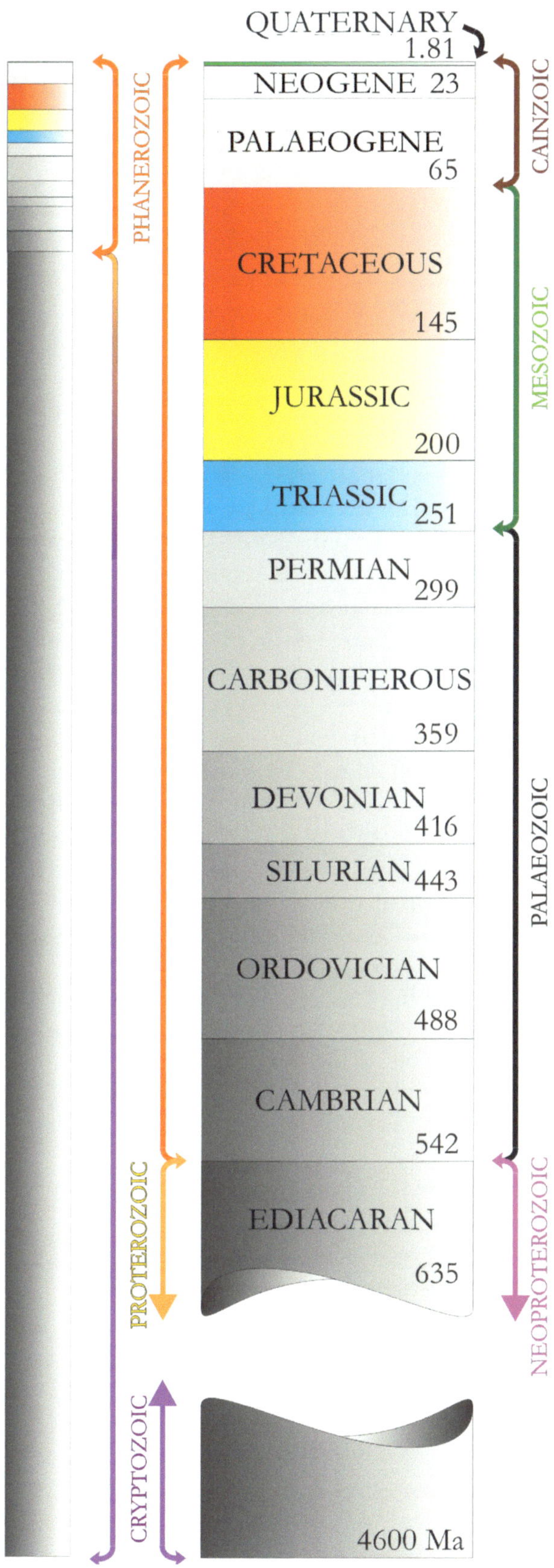

The geological time-scale.

The geological time-scale shows the relative time ranges (left) of the Cryptozoic (4600–2500 million years ago; purple line), Proterozoic (2500–1000 million years ago) and Phanerozoic (542 million years ago to the present; orange line) eons. The various eras and periods of the Phanerozoic are also depicted with their respective time ranges on the right.

The Cainozoic Era (brown line) comprises the Palaeogene (65–23 million years ago), Neogene (23–~1.81 million years ago) and Quaternary (~1.81 million years ago to present) periods and is often informally called the 'Age of Mammals'.

The Mesozoic Era (green line), or 'Age of Dinosaurs', is the principal focus of this book and is therefore highlighted. It includes three periods: the Triassic (blue), which extended from ~251–200 million years ago, the Jurassic (yellow) from 200–145 million years ago and the Cretaceous (red), the longest of the three periods, extending from 145–65 million years ago when the dinosaurs and many other animals and plants became extinct.

The Palaeozoic Era (black line) marks the beginning of the Phanerozoic. It records the explosive development of the earliest animal life during the Cambrian (542 million years ago) through to the Permian (299–251 million years ago), which terminates with the largest extinction event of all time – around 80% of life disappeared.

The start of the Cryptozoic Eon is marked by the formation of the Earth and extends through the appearance of the first single-celled organisms ~3.5 billion years ago. Towards the end of the Proterozoic (a time-frame also known as the Neoproterozoic), complex organisms first started to appear (Ediacaran, 635 million years ago).

provide a clear picture of how and, importantly, when large-scale changes in the Earth's flora and faunas have occurred throughout geological time.

The earliest evidence for life on Earth is known from rocks ~3.5 billion years old – a time-frame known as the Cryptozoic Eon ('hidden life'). Most conspicuous are fossil stromatolites, domed limestone structures produced by cyanobacteria. Cyanobacteria are photosynthesisers, using the light-harvesting green pigment chlorophyll to manufacture complex sugars as food. Stromatolite fossils do not represent the organisms themselves; rather, they are an accreted structure formed from trapped silt particles and lime precipitate by the encrusting bacteria. Stromatolites created massive reefs during the later Cryptozoic (termed the Proterozoic), long before modern reef-builders such as corals had evolved. The Cryptozoic also marks the time when the first bacteria (prokaryotes – cells lacking a nucleus), fungi and single-celled eukaryotic (nucleus-bearing) organisms appeared, groups that ultimately gave rise to the diverse range of living organisms that we see today.

The activities of early photosynthesising organisms such as cyanobacteria are the source of our modern oxygen-rich atmosphere. Their production of carbohydrates from carbon dioxide (CO_2) and the release of hydrogen from water (H_2O) have yielded oxygen (O_2) as a by-product. Over time, the increasing oxygen levels prompted a rise in the density of the ozone layer. This ultimately provided an effective filter against the ultraviolet radiation in sunlight and thus paved the way for the evolution of oxygen-utilising organisms.

Around 2.5 billion years ago atmospheric oxygen reached a level sufficient for oxidisation of sulphur and iron in the global oceans. This led to the deposition of banded iron formations, which provide most of the world's iron ore. It also marks the beginning of the Proterozoic Eon ('earlier life'). The transition to an oxygen-rich atmosphere continued through the Proterozoic and was accompanied by several epic glaciations during the Neoproterozoic Era ('new-earlier life' – the youngest part of the Proterozoic, previously known as the Precambrian), including a near-total freeze of the Earth in the Cryogenian (850–635 million years ago). Evidence for glacial build-up has been found even at the ancient equator. In the wake of this climatic extreme the first multicellular (metazoan) organisms appeared, ~635 million years ago. These 'true animals' were mainly confined to the oceans, and would be for the next 200 million years. An incredible array of early multicellular creatures has been found in Australia, dating to ~560 million years ago. These are collectively called the Ediacaran Biota (named after Ediacara, South Australia, where they were first discovered). They include some of the earliest known soft-bodied (lacking hard skeletal elements) ancestors of modern animals such as cnidarians (jellyfish and corals), annelids (worms) and possibly even chordates (the group that includes vertebrates). By ~542 million years ago all the Ediacaran animals had disappeared, marking the end of the Ediacaran Period. They were replaced by a rapid diversification of later organisms, many of which had hard parts incorporated into the outer surfaces of their bodies. This appearance of hard shells was a landmark in the history of life on Earth and is evidenced by a sudden increase in the range of animals preserved in the fossil record. Palaeontologists have called this event the 'Cambrian explosion', referring to the Cambrian Period (542–488 million years ago), during which it occurs, and the manner in which many kinds of hard-shelled fossil animals (e.g. trilobites) suddenly appeared. The Cambrian also marks the beginning of the Palaeozoic Era ('ancient life'), the first major subdivision of the Phanerozoic Eon ('visible life'). It incorporates five other periods:

- Ordovician (488–443 million years ago), marked by the diversification of the earliest vertebrates;
- Silurian (443–416 million years ago), when the first land plants evolved;
- Devonian (416–359 million years ago), where fish diversified and the first land animals appeared;
- Carboniferous (359–299 million years ago), which is marked by the earliest reptiles and the spread of life onto land;
- Permian (299–251 million years ago), the terminus of which records the greatest extinction event of all time, involving ~80% of known species.

Examples of Ediacaran fossils from the Flinders Ranges in South Australia. (Clockwise from top left) The pennatulate (soft coral) *Charniodiscus*, a putative early chordate, the cnidarian *Mawsonites*, a possible early trilobite *Parvancorina*, an early edrioasteroid (echinoderm) *Tribrachidium* and a flat-bodied worm (annellid) *Dickinsonia*. Images courtesy James Gehling.

Some of the most characteristic Cambrian fossils are trilobites. These early arthropods were conspicuous elements in marine faunas until the end of the Palaeozoic Era. *Lyriaspis sigillum* depicted is from the middle Cambrian Beetle Creek of western Queensland.

The massive extinction at the end of the Palaeozoic provided an opportunity for new species to evolve. These filled the ecological vacuum left in many terrestrial and marine environments. The ancestors of most modern marine invertebrate lineages appeared at this time. An important example is bivalve molluscs, a group that rapidly diversified after the Permian to become a dominant element in benthic (bottom-dwelling) invertebrate faunas throughout the Mesozoic Era ('middle life'). The Mesozoic began ~251 million years ago and is more popularly known as the Age of Dinosaurs; it will be discussed in more detail in Chapter 2. Importantly, the Mesozoic saw the appearance of many animals and plants that are still extant today. Mammals, birds, crocodiles, turtles and flowers all originated alongside the dinosaurs but managed to survive the mass extinction at the end of the era. Often called the Cretaceous-Tertiary (K-T) or more correctly the Cretaceous-Palaeogene (K-P) boundary event, the terminal Mesozoic extinction is dated at ~65 million years ago. It recorded the disappearance of dominant Mesozoic life forms such as dinosaurs, most marine reptiles, pterosaurs and ammonites. Causes of the K-P event are highly speculative but they may include factors ranging from long-term ecosystem change to dramatic asteroid impacts. Some evidence exists to support both models, but it is likely that a combination of factors placed stress on environments at the end of the Mesozoic and ultimately facilitated the rapid extinction of species unable to adapt to their changing ecosystems.

Archaeocyathans were some of the earliest reef-forming animals. These sponge-like organisms used a porous skeleton to sieve tiny food particles from the water. (Left) Cross-section through an archaeocyathid *Metaldetes* from the Lower Cambrian of South Australia. (Right) *Archaeocyathus* from an Early Cambrian patch reef in the Flinders Ranges, South Australia.

Following the removal of reptiles from most major niches, mammals became the dominant vertebrates in many terrestrial habitats. This transition is encapsulated by the Cainozoic Era ('recent life') – the Age of Mammals – and represents the time-frame in which we live. The bulk of the Cainozoic involved the Palaeogene (65–23 million years ago) and Neogene (23–1.81 million years ago), periods usually described in terms of their five consecutive epochs:

- Paleocene (65–55.8 million years ago), recording the first diversification of Cainozoic mammals;

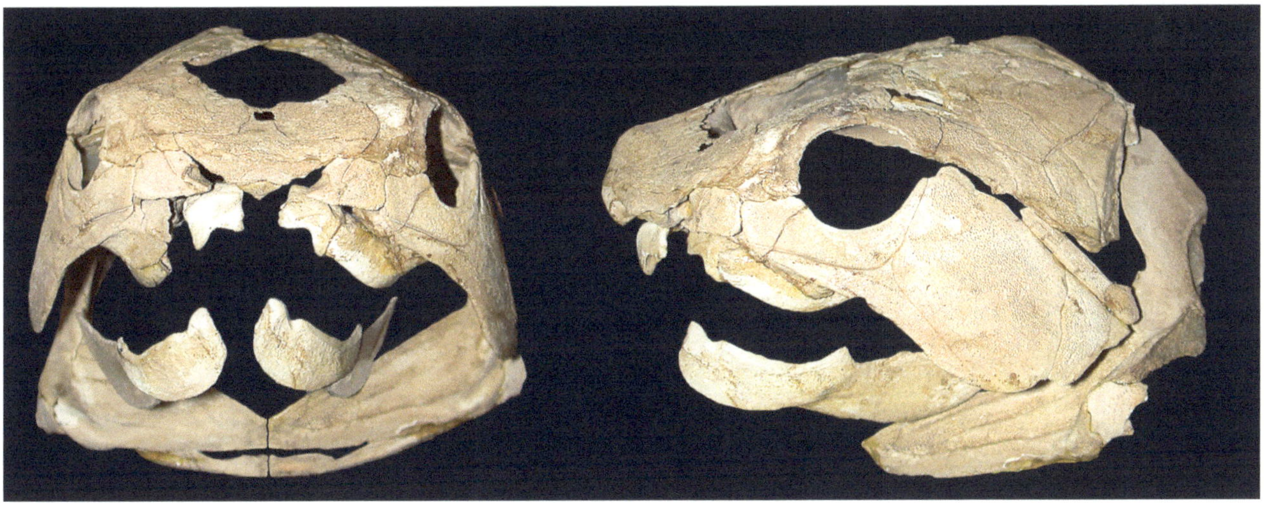

During the Devonian, fish diversified into many marine and freshwater environments. One of the most common ancient groups were placoderms. These had characteristic armour-plated skulls and were both bottom-feeders and top-level pelagic predators. This example, *Eastmanosteus*, is from the famous Gogo locality in Western Australia, a Devonian reef deposit that has yielded evidence of a complex fish community inhabiting the north-western margin of the Australian continent ~380 million years ago.

Early plants such as *Baragwanathia* (a type of clubmoss or lycopod) are thought to have proliferated along the margins of watercourses during the Silurian. By the late Palaeozoic, plants dominated both wetland and dry upland environments. This single *Baragwanathia* frond is from Mudgee in New South Wales.

- Eocene (55.8–33.9 million years ago), including the earliest whales and songbirds;
- Oligocene (33.9–23.03 million years ago), which saw the first appearance of kangaroos in Australia;
- Miocene (23.03–5.33 million years ago), during which the ancestors of Australia's modern arid-adapted marsupial lineages began to diversify;
- Pliocene (5.33–~1.81 million years ago), which saw the radiation of hominids including our species, *Homo sapiens*.

The Quaternary began ~1.81 million years ago and is our own period of geological time. It records the great ice ages of the Pleistocene Epoch, which ended ~10 000 years ago, and witnessed the decline of large mammals across the globe (e.g. mammoths and sabretooth cats in the northern hemisphere and giant marsupials in Australia). The extinction of the Pleistocene megafauna has been variously linked to climate change and/or ecological destabilisation triggered by human hunting. Whatever the causes, the rapid spread of our species is the key event marking the current Holocene Epoch, which incorporates the dawn of civilisation and the genesis of our modern technology-based society.

The precise dating of rocks

The absolute dating (physical time dating) of rocks was developed relatively recently, in the mid 20th century. Most techniques rely on radioactive processes that proceed at a constant rate regardless of chemical, temperature and pressure conditions in the Earth's crust. Radioactive elements are known to decay over a set period called a 'half-life'. During this time the original amount of radioactive material is reduced by one-half; it continues halving at the same rate until the measurable radioactivity becomes undetectable relative to the background radiation from space.

Radiometric (measuring radioactivity) dating procedures can utilise different radioactive materials, all with different properties and half-lives. One common element is carbon 14, often used for dating charcoal and plant material found in sites no older than about 35 000 years (half-life ~5700 years). Carbon 14 is a radioactive isotope (or form) of carbon, characterised by 14 particles making up the nucleus of the atom. It is constantly produced by radioactive bombardment of the atmosphere from space and forms part of the air taken in by every living organism. Carbon 14 is also combined into plant tissues and is therefore often consumed by animals. The naturally occurring atmospheric ratio

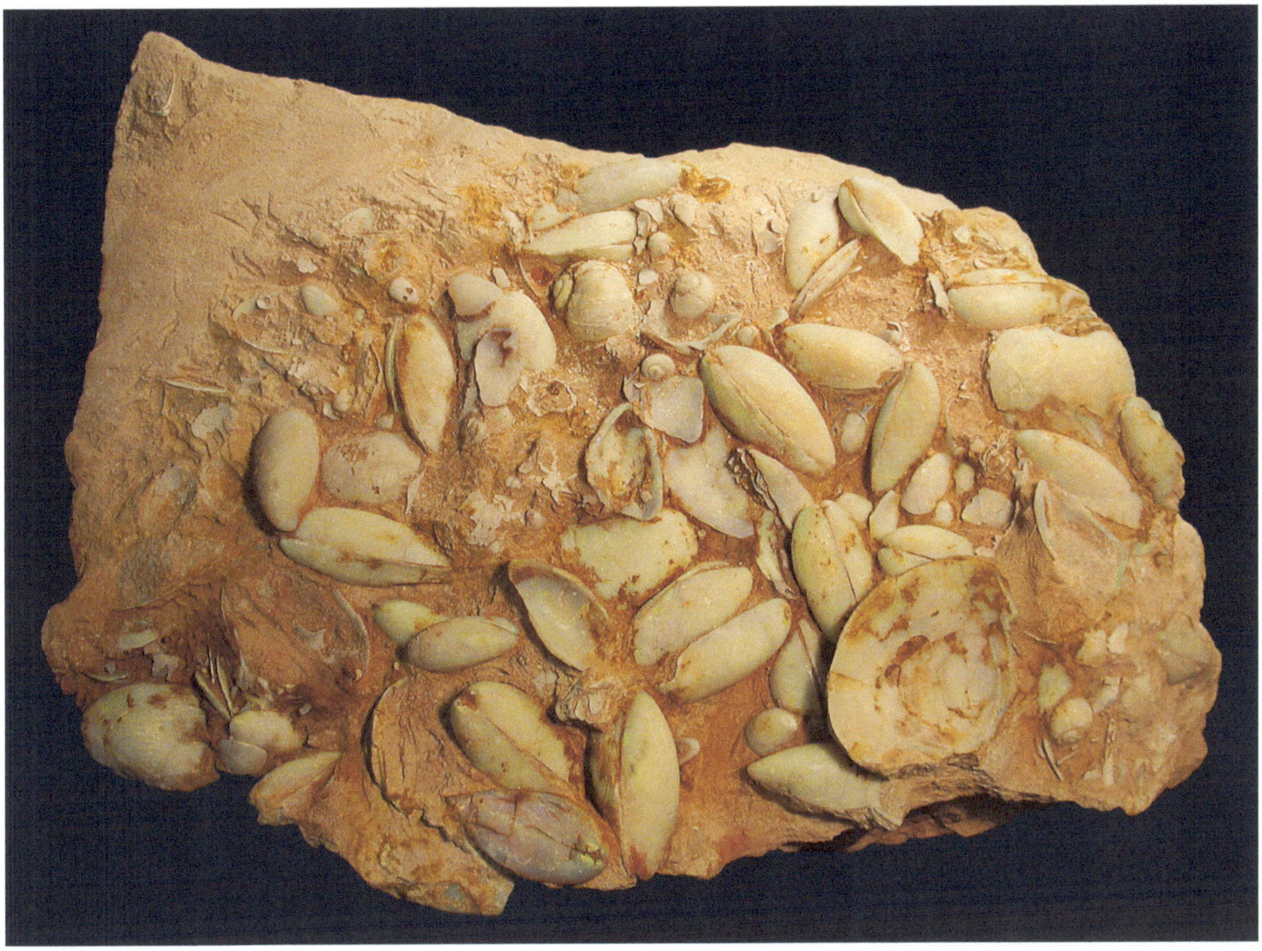

Bivalves are one of the most ubiquitous fossils in Mesozoic and Cainozoic sediments. They have dominated marine and freshwater benthic environments since the end of the Palaeozoic. This collection of opalised bivalve (including the mussel-like *Eyrena* and scallop-like *Maccoyella*) and snail (*Euspira*) shells is from the Cretaceous deposits at Coober Pedy in northern South Australia.

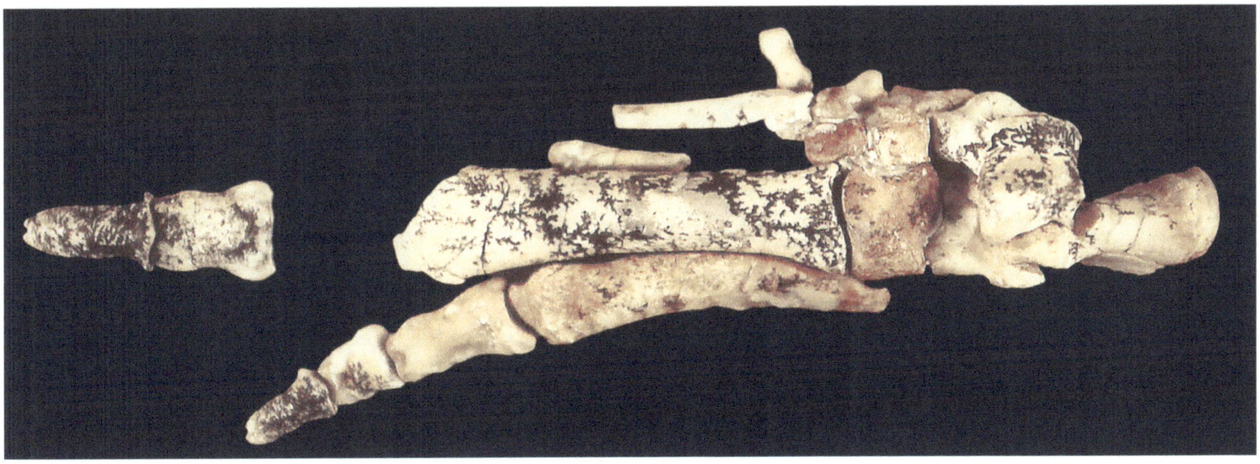

During the Cainozoic, Australian terrestrial faunas were dominated by marsupials. This articulated foot is from one of the earliest known fossil kangaroos. Named *Nambaroo*, this early macropod was only distantly related to modern kangaroos. It probably bounded on all fours rather than hopped.

of carbon 14 relative to its non-radioactive lighter isotopes carbon 12 and carbon 13 is known, and remains constant in organisms throughout their lives. After death, carbon 14 is no longer incorporated into the tissues and any previously absorbed amounts begin to decay. By measuring the ratio of carbon 14 relative to carbon 12 and carbon 13 in fossil material, the amount of carbon 14 decay can be determined, and thus the amount of time lapsed between the organism's death and the present day is pinpointed. This process is called carbon dating.

Potassium-argon is another radioactive dating technique used for samples ranging in age from ~100 000 years to 4 billion years. This method utilises isotopes of potassium 40, which decay to argon 40 over a half-life of ~1.25 billion years. The decay process is known as electron capture; it involves the transformation of the alkali metal potassium 40 into the non-reactive gas argon 40 via a change in atomic structure. The potassium 40 is physically trapped in volcanic rocks when it is produced as molten lava from volcanic vents and fissures. When the lava cools and hardens, the potassium 40 accumulates in crystals and begins to decay to argon 40, thus starting the clock.

Because potassium-argon dating is reliant upon volcanic rocks, it cannot be applied directly to the fossil material preserved in sedimentary strata. Nevertheless, it can be a source of very reliable dates for volcanic deposits (commonly volcanic ash) that are interspersed with sedimentary layers. For example, a sedimentary layer containing fossils that is sandwiched between volcanic rocks can be given an age range based on the younger volcanic material above (providing an uppermost date) and the older volcanic material below (providing a lowermost age).

The famous Dinosaur Cove site (with fieldwork participants highlighted for scale) on the south coast of Victoria has produced a diverse assemblage of Early Cretaceous vertebrates. Age determination of the specific sites has been undertaken using various methods including fission track dating. This has made it one of the best-dated Mesozoic fossil localities in Australia. Image courtesy Thomas Rich.

This utilises the concept of 'superposition', a scientific law that considers the lowest rock layers to be the oldest (barring any disturbance by tectonic activity). Fossils from the sandwiched sedimentary layer may then be used as 'index' fossils (species with a wide geographic and ecological range but short-lived species life-span). These can be correlated with similar fossil types in non-volcanically constrained sedimentary sequences elsewhere to provide a reasonable age limit.

Another radioactive dating method utilising volcanic rocks is fission track dating. This relies upon the spontaneous splitting or fission of uranium 238 or uranium 235 isotopes, which occurs at a constant rate through time. The dense nucleus of the uranium 238/uranium 235 atom is known to split into two fragments that fly apart at high speed and leave visible damage trails or 'fission tracks' in any enclosing material (usually crystals). The frequency of these splits increases as the isotopes decay over time. Consequently the number of trails, together with the known half-life of the uranium isotope, can provide a measure of the age of the enclosing structure. Fission track dating is used for minerals such as apatite, zircon and some volcanic glasses (e.g. obsidian), and has been successfully applied to sedimentary sequences between ~20 million and 1 billion years old. Ideally, the fission track technique is applied as a comparative test for other dating methods, particularly potassium-argon. Notably, all rock dating is based upon multiple lines of evidence and thus the ages given for fossils, although imprecise when considered in terms of single years, are nonetheless extremely reliable with respect to the vastness of geological time.

The mobile crust: plate tectonics

It is important to remember that the Earth's continents and oceans are not static but have changed shape and position constantly throughout geological time. The movements and rearrangements of landmasses are explained by the theory of plate tectonics.

The Earth's crust is not a continuous sheet; rather, it is made up of a series of irregularly shaped lithospheric plates upon which ride the continents and ocean floor. Convection currents in the hot molten mantle (under the crust) drive the plate movements, and the boundaries between plates are active zones where earthquakes, mountain-building and volcanic activity occur. There are four major types of plate boundary:

- spreading zones (mid-oceanic ridges), where new crust is produced and plates move apart;

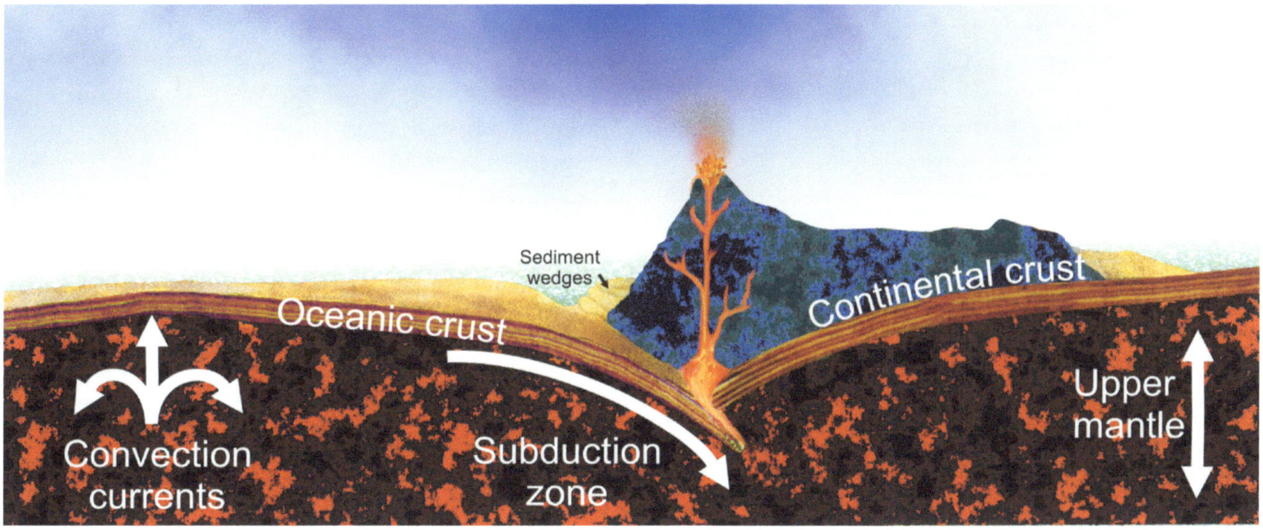

A subduction zone where convergent plates driven by convection currents override one another, forcing plate material down into the mantle to melt and rise again to the surface. Subduction processes are accompanied by buckling and uplift of the crust at plate boundaries, to form high volcanic mountain chains.

- subduction zones, where plates converge and one overrides the other. The underlying or subducted plate is forced down into the mantle where its rocks melt and the less dense magma (molten rock) rises to the surface through volcanoes;
- collision zones, where continents riding on the plates are colliding;
- transform faults, where two plates slip past each other without the destruction or production of plate material.

The past orientations of crustal plates and their continental passengers can be determined through various lines of evidence. Palaeomagnetism (the study of the Earth's magnetic field through minerals in rocks) is a means of determining continental orientation relative to ancient north and south magnetic poles. It may be surprising to learn that the north and south poles have swapped magnetic polarity numerous times throughout the Earth's history. The record of these shifts is preserved in magnetic minerals trapped in rocks. By knowing the age of these rocks a magnetic log can be compiled, thus mapping the changing orientation of the continents through time.

Information from palaeomagnetism is usually coupled with a wealth of other data including information from fossils. For example, the distributions of key fossil types such as the Permian plant *Glossopteris* and the aquatic reptile *Mesosaurus* show that the African and South American landmasses, today separated by the Pacific Ocean, must once have been joined. *Glossopteris* was a large tree-like plant, fossils of which are found in Australia, India, Antarctica, southern Africa and South America. These continents are now separated by oceanic barriers and could not be easily traversed by a stationary plant. Similarly, the small (300 mm) reptile *Mesosaurus* is unlikely to have migrated vast distances across open ocean, thus the recovery of its fossils from rocks of the same age in eastern South America (Brazil) and southern Africa (South Africa) strongly suggests that these two continents must been juxtaposed in the past.

Continental dynamics and positioning during the Cryptozoic and Proterozoic are poorly known. In contrast, movements in the Phanerozoic are based on a wealth of readily accessible geological data.

Early in the Phanerozoic (Cambrian, ~540 million years ago) most of the Earth's landmasses were bound into a southern supercontinent called Gondwana, with Australia and Antarctica at the equator. A number of smaller continental fragments were also arranged along the equator, including Laurentia (now part of North America), Siberia, Kazakstania (now Kazakhstan), Baltica (present-day Poland and the Baltic states) and parts of China. By ~300 million years ago (Late Carboniferous), these landmasses accreted with Gondwana to form a single massive supercontinent called Pangaea. This stretched from pole to pole and opposed a gigantic ocean, Panthalassa. Around 250 million years ago (Triassic), Pangaea began to break up, forming two smaller supercontinents: Laurasia (North America, Europe and most of Asia) in the northern hemisphere and Gondwana (Africa, Madagascar, South America, India, Antarctica, Australia, Turkey, the Arabian Peninsula and parts of South-East Asia) in the southern hemisphere. These were still connected by a land bridge through northern Africa, but most of their area was separated by the vast Tethys Ocean.

About 360 million years ago, Australia shifted dramatically from its previously equatorial position towards the South Pole. It occupied this southern high-latitude region for about the next 270 million years. However, cold conditions, ice sheets and an absence of vegetation did not always characterise the ancient Australian continent. Indeed, throughout much of the Mesozoic, global temperatures were much warmer than they are today, although at certain times such as the late Palaeozoic (Permian, ~260 million years ago) and Mesozoic (Early Cretaceous, ~120 million years ago), distinct cold-snaps are recorded in the Australian high-latitude sediments.

By the late Mesozoic (Late Cretaceous, 96 million years ago) Gondwana had begun to break up. Africa and South America were separated by the infant Atlantic Ocean, India had broken away and was slowly drifting north, New Zealand had become a series of isolated islands, and the first sea floor was being laid down between Antarctica and Australia. By the middle Palaeogene (Eocene, 49 million

Fossils of the Permian land plant *Glossopteris* (top) and the small aquatic reptile *Mesosaurus* (bottom) have been found in similar-aged deposits across the fragmented remnant landmasses of the ancient Gondwanan supercontinent. Such transcontinental distributions have been widely cited as evidence for movement of the crustal plates.

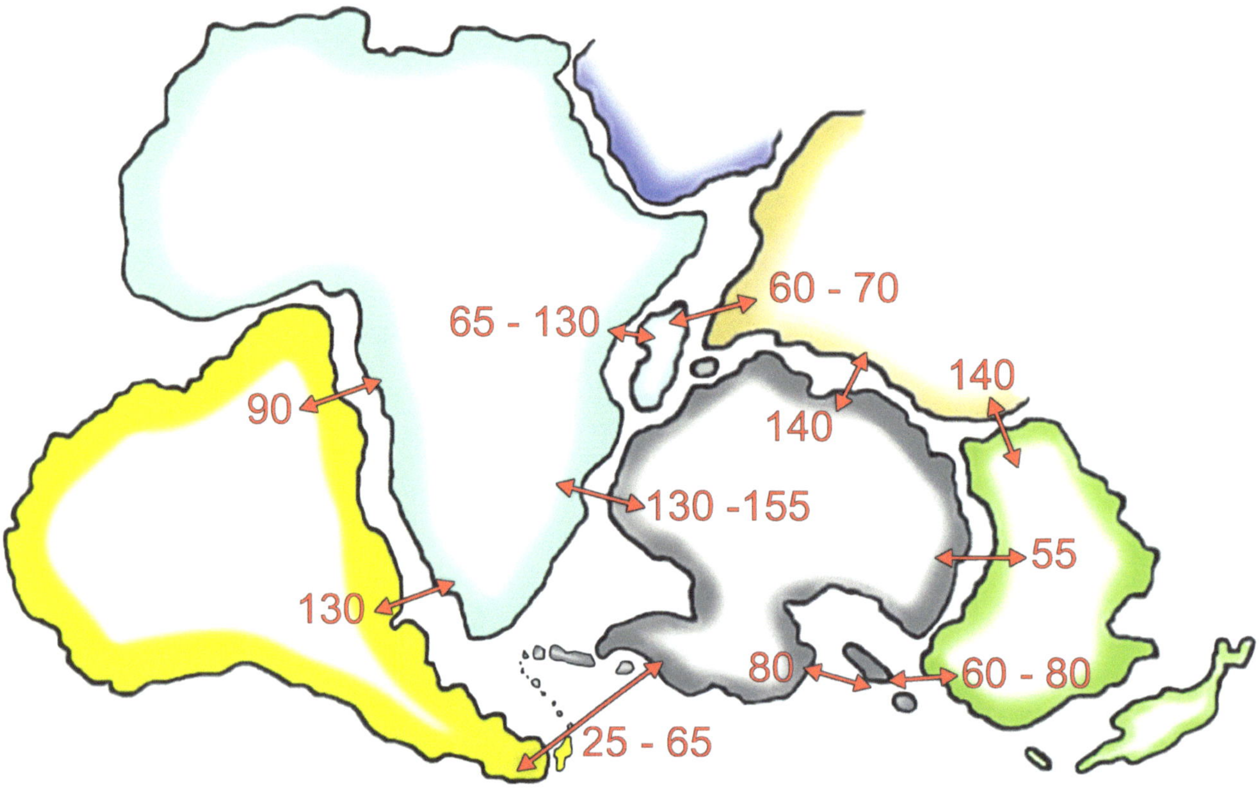

A basic map showing the divergence times (in millions of years) of the principal Gondwanan landmasses. Modern continents include Africa and Madagascar (aqua), Antarctica (grey), Arabia (blue), Australia (green), India (brown) and South America (yellow).

years ago) the split was complete. Australia had become the island continent and over the next 50 million years it drifted slowly north, with its cargo of unique animals and plants.

Finding, preparing and interpreting fossils

Fossils occur surprisingly frequently and can be found by anyone who knows where to look. The trick of unearthing a good specimen relies on a variety of factors, not the least being serendipity. Great fossil discoveries have been made in exotic locations, dusty museum drawers and even city suburbs. The people who have made them include palaeontologists, keen-eyed enthusiasts, exploration geologists mapping for oil or minerals, opal miners, farmers or just people out for a walk.

The easiest way to find fossils is to start with a geological map that shows where sedimentary rocks of the right age and type might occur. By surveying these exposures (outcrops), you may be lucky enough to turn up a good specimen. Exposures may produce prolific amounts of material, or none. The search can be refined by reading geological reports on the target areas. Geological reports are usually compiled by exploration geologists looking for ore deposits or drilling for oil and gas. If fossils are found in the course of this work, their presence is recorded and a reference sample collected for storage in a geological survey or museum collection. Examining these specimens can provide valuable leads on where to find new fossil sites. Specimens can be held in university, museum or private collections for years without their scientific importance being recognised. Some may represent important discoveries, and other fragmentary fossils might lead to additional finds because of their locality data.

When a fossil is located in the field, it can be difficult to extract it without damage. Most fossils are extremely fragile and can be cracked or reduced to little more than a powdery residue by surface weathering or the actions of plant roots. Careful

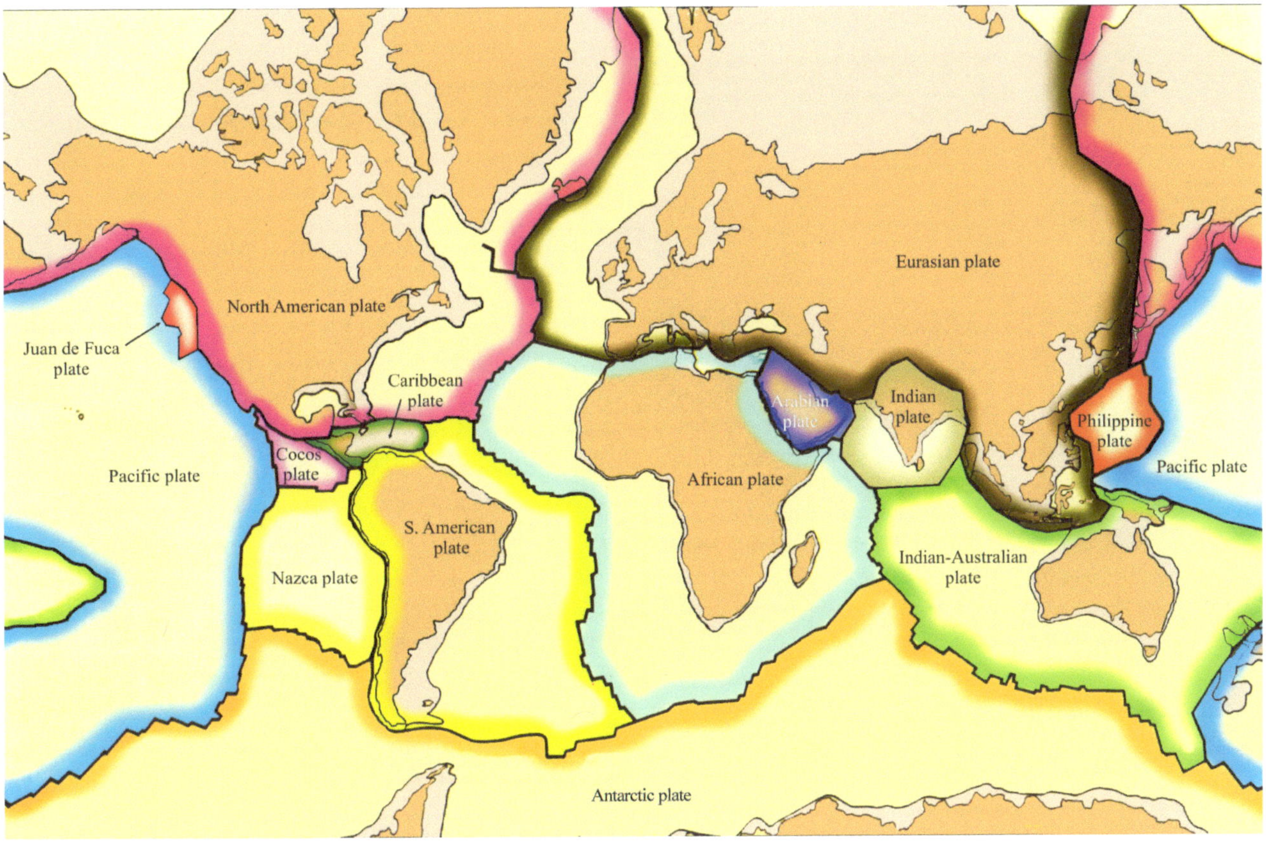

The modern continental distribution relative to crustal plates and plate boundaries.

collection and expert preparation in the laboratory is critical for the recovery of an intact specimen.

The standard techniques for extracting fossils in the field have not changed for over 100 years. Nevertheless, each new specimen presents a unique set of problems that often require modern solutions and some ingenuity. A good example is collecting large vertebrate skeletons. These are usually recognised by a trace of bone on the surface or perhaps in the wall of an opal mine. After an initial assessment, the surrounding matrix (encasing rock) must be carefully excavated (by hand or with earth-moving machinery) to expose the rest of the fragile skeleton. During this process the newly revealed bones are consolidated using glues to prevent any breakage. Once fully uncovered, the skeleton can be coated in a protective wrapping of hessian or fibreglass soaked in plaster. This is applied over a separating layer of wet paper to stop plaster adhering to the bone, and eventually forms a hard jacket. This holds the specimen together and keeps the bones protected in their original position for the journey back to the laboratory. Once there, the jacket can be opened to allow further preparation and complete extraction of the specimen.

The preparation techniques for fossils depend on the type of rock matrix and composition of the fossil remains. Soft shales, mudstones and silicified (turned into silica) sediments are usually removed with mechanical tools such as hand-held rotary grinders and pneumatic scribes. If fossils are preserved in soft rocks they can be worked with air abrasive units, which use fine powders shot at high pressure by compressed air to remove matrix. When fossils are preserved in limestone they can be treated with acid. Limestone is composed of calcium carbonate ($CaCO_3$) and is chemically different from the common minerals comprising fossil bone (calcium phosphate, $Ca_3[PO_4]_2$) or silicified remains (usually basic silica [SiO_2] plus impurities).

Plaster jacketing is a standard technique for securing and packaging large fragile fossil specimens in the field. The process involves application of a layer of plaster-soaked hessian or fibreglass strips to the surface of the fossil, with a separating layer of wet paper to prevent sticking. When this hardens it forms a tough protective jacket.

It therefore reacts readily with low concentrations of dilute acetic, formic or hydrochloric acid. The fossils remain unaffected. The acid dissolution process is slow, taking many months or even years to completely free the fossil from its limestone matrix. However, the results can be spectacular, revealing full surface texturing and even fine internal structures preserved in three-dimensional detail.

The final step in the process – interpreting the fossil – is often the hardest because weathering, distortion by earth movements or fragmentation can render some structures virtually impossible to identify with certainty. The best tool to begin with is comparative anatomy, the process of comparing the fossil with other specimens of living and extinct animals and plants in the hope of finding a match. The published scientific literature can be invaluable, providing descriptions and photographs of remains held in collections across the world. The key is to look for common diagnostic characters. These are particular features that characterise every type of animal and plant and provide the basis for modern systematics (the study of the identification and classification of organisms).

Determining the relationships of organisms

An important use of fossils is in the reconstruction of evolutionary relationships. One of the most common methods for doing this is cladistics, which utilises specifically identified derived characteristics (synapomorphies) to establish homology or common descent between different types of organisms. Synapomorphic characters can be extracted from anatomical, behavioural, physiological, molecular or other data, and are found in fossil and living organisms. They are best identified on the basis of close structural similarity or a common developmental pathway and need to be congruent with other synapomorphies supporting the same grouping. Synapomorphic characters are used to define individual organisms, collectively termed 'taxa' (singular = taxon), and arrange them into groups of common ancestry called clades. For example, as an ancestral species evolves into one or more descendent species, its original characteristics are inherited by its offspring. These characteristics are thus shared by the ancestor species and all its descendants, and become synapomorphies for that collective clade. Consequently, the distribution of various synapomorphies among the wide range of living and extinct organisms can be used to determine both the relationships of different clades and the sequences in which they evolved. The patterns of evolution reconstructed using cladistic analyses are usually depicted as cladograms, branching diagrams that illustrate the common ancestry in interspecies relationships. They provide a basis for establishing sister species – taxa that share a most recent common ancestor. It should be noted that sister species relationships are based purely on the

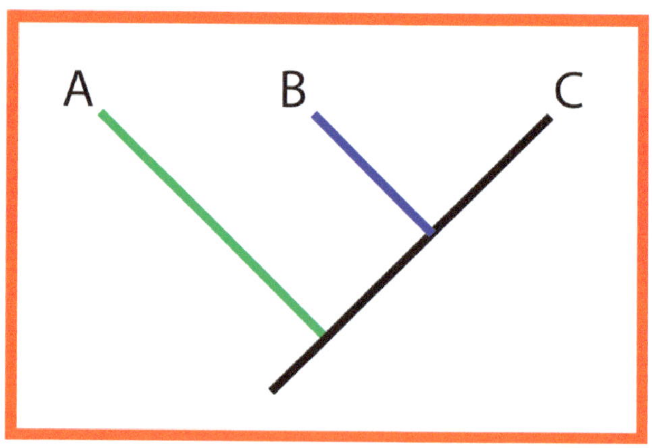

A simple cladogram depicting the relationships of three taxa (A, B, C). In this arrangement B and C are shown to be more closely related than A.

synapomorphies present in the constituent taxa and do not necessarily take the relative ages of those taxa into account.

Reconstructing the evolutionary genealogy (phylogeny) of a species or clade is often complicated because characters do not always evolve only once or in a linear pattern. Indeed, in some cases similar characteristics may evolve in organisms that share no common ancestry at all – this is termed 'convergent evolution'. The phenomenon of convergent evolution seems to have been ubiquitous throughout the history of life, and is often attributed to organisms adapting to the common physical constraints of the Earth's natural environments. A good example is the evolution of wings for flight. This lifestyle has very specific constraints involving effective movement through a gaseous medium, and thus has prompted the independent development of comparable stuctures in insects, reptiles (pterosaurs), birds and mammals (bats).

Taking aspects such as convergent evolution into account, the critical first step in reconstructing the phylogenetic relationships of any species or clade is to determine whether the various characters are likely to have evolved from a common ancestor or from separate structural origins. The most widely accepted method for examining this is the principle of parsimony. Parsimony relies on the idea that evolution always follows the simplest path, or that the hypothetical arrangement of species or clades in a cladogram supported by the greatest amount of available evidence (and contradicted by the least) should thus be the most plausible. Parsimony is applied in a modern cladistic analysis by comparing large numbers of characters from many species or clades, with preliminary hypotheses of homology, to determine whether a character is pre-existing ('primitive') or has been transformed through evolution ('derived') from a pre-existing form. A numerical value is assigned to each characteristic and a numerical matrix is compiled. The matrix is analysed using computer algorithms designed to search for combinations of similar values that correspond to the most parsimonious hypothesis. The resulting cladogram summarises the most parsimonious relationships, based on the available data (the largest number of derived similarities or synapomorphies), and can provide a testable foundation for determining instances of common ancestry and convergent evolution.

2

SETTING THE STAGE: THE MESOZOIC ERA

The Mesozoic Era – the Age of Dinosaurs – began ~251 million years ago and ended ~65 million years ago. Its 180 million-year timespan is subdivided into three geological periods, each identified by a distinctive suite of fossils. The oldest of the three periods, the Triassic (251–199.6 million years ago), is characterised by the radiation of the earliest modern marine invertebrates. It saw the rapid diversification of bivalve molluscs, sea urchins, starfish and advanced colonial corals. Cephalopods (squid, octopus etc.) underwent a major increase in variety at that time, with the first appearance of ammonoids and belemnoids as dominant elements in many pelagic (open-water living) marine invertebrate faunas. Marine vertebrate assemblages included primitive bony fish, modern sharks and marine reptiles such as ichthyosaurs (dolphin-like marine reptiles) and sauropterygians, the group that includes plesiosaurs. On land, giant crocodile-like temnospondyl amphibians, mammal-like reptiles (therapsids) and early archosauromorphs (ancestors of crocodiles and dinosaurs) were the dominant components of vertebrate assemblages during the Early to Middle Triassic. However, by the Late Triassic these were replaced by new forms including the earliest true mammals, turtles, sphenodontids (Tuatara), crocodiles, pterosaurs (flying reptiles) and dinosaurs. Land vegetation during the Triassic was rich in clubmosses, podocarp conifers (primitive pines), ginkgos or maidenhair trees and cycads. Mosses and liverworts were also present; flowering plants had not yet evolved.

The succeeding Jurassic (199.6–145.5 million years ago) was a time of globally warm and wet conditions. These prompted the development of a uniform cosmopolitan flora of conifers, cycads, ferns, seed-ferns (primitive fern-like plants), ginkgos and clubmosses. This distinctive vegetation persisted for the next 70 million years and was the last to be composed of plants from ancient groups. Feeding on this rich plant-life was the first great radiation of dinosaurs, which by the Early Jurassic had replaced the archetypal Triassic mammal-like reptiles and primitive archosauromorphs as the dominant large terrestrial (land-based) vertebrate lineage. Prominent Jurassic dinosaur groups included the sauropods (*Brontosaurus*-like dinosaurs) and armoured stegosaurs. Ornithopods (bipedal herbivorous dinosaurs) and theropods (carnivorous dinosaurs) were also common. The latter gave rise to birds; the earliest known definitive bird fossils have been recovered from Upper Jurassic deposits in Germany (*Archaeopterix*).

Life in the Jurassic oceans was somewhat different from that of the Triassic. A major extinction had reorganised marine invertebrate faunas, allowing ammonites and bivalves to emerge as dominant surviving components. Marine reptiles also underwent an extinction event at the end of the Triassic: basal sauropterygians disappeared and were replaced by plesiosaurs, marine crocodiles and ichthyosaurs.

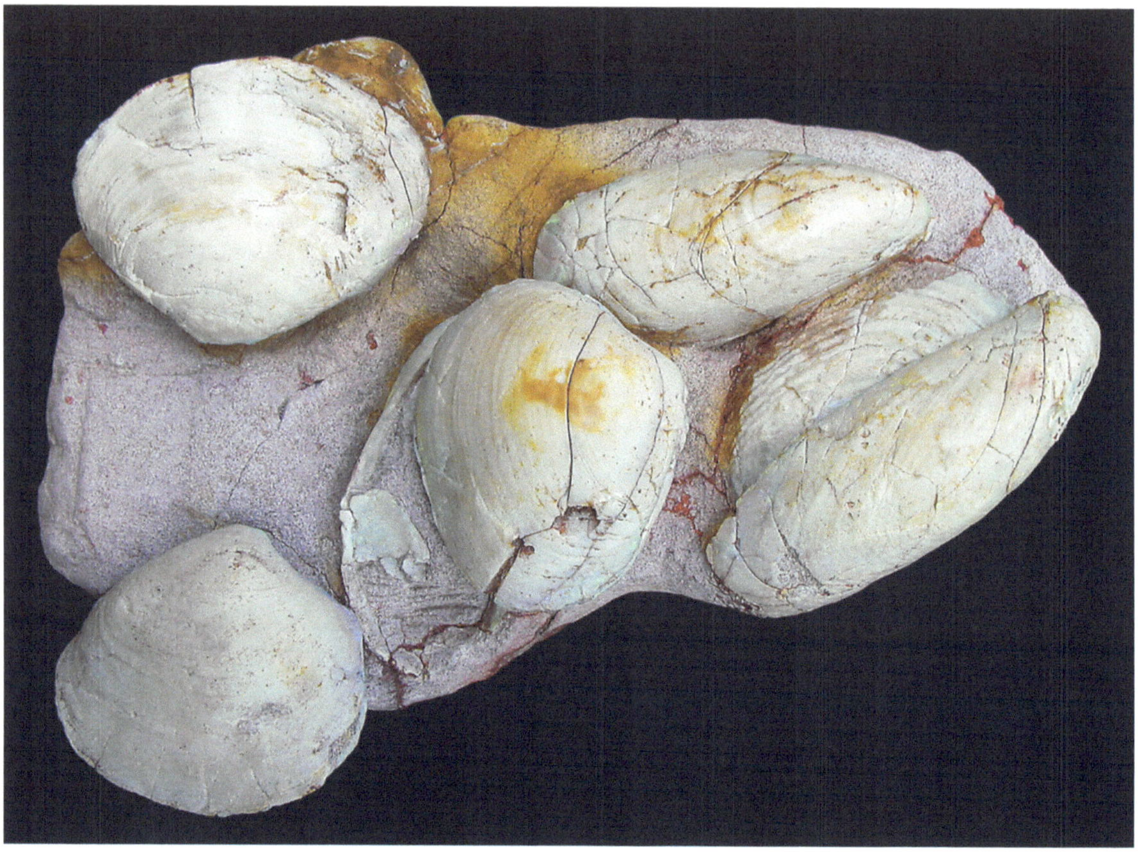

Bivalves are one of the most ubiquitous groups of modern marine invertebrates. Their remains are extremely common in Australian Mesozoic strata, including the opal-bearing sediments at Coober Pedy in South Australia. The shell accumulation shown here incorporates several different taxa. (Clockwise from top) The cockle-like *Cyrenopsis australiensis*, the mussel *Eyrena tatei* and the partially exposed ribbed shell of a burrowing trigoniid *Laevitrigonia*.

The maidenhair tree or *Ginkgo biloba* is a common plant in many gardens. The species is a living fossil and its record stretches back to the Mesozoic, over 200 million years ago.

ERA	SYSTEM	SERIES		STAGE	Ma
MESOZOIC (Mz)	CRETACEOUS (K)	LATE (K_2)	Senonian	Maastrichtian	65.5
				Campanian	70.6
				Santonian	83.5
				Coniacian	85.8
			Gallic	Turonian	88.6
				Cenomanian	93.6
		EARLY (K_1)		Albian	99.6
				Aptian	112.0
				Barremian	125.0
			Neocomian	Hauterivian	130.0
				Valanginian	133.9
				Berriasian	140.2
	JURASSIC (J)	LATE (J_3)	Malm	Tithonian	145.5
				Kimmeridgian	150.8
				Oxfordian	155.6
		MIDDLE (J_2)	Dogger	Callovian	161.2
				Bathonian	164.7
				Bajocian	167.7
				Aalenian	171.6
		EARLY (J_1)	Liassic	Toarcian	175.6
				Pliensbachian	183.0
				Sinemurian	189.6
				Hettangian	196.5
	TRIASSIC (Tr)	LATE (Tr_3)		Rhaetian	199.6
				Norian	203.6
				Carnian	216.5
		MIDDLE (Tr_2)		Ladinian	228.7
				Anisian	237.0
		EARLY (Tr_1)		Olenekian	245.9
				Induan	249.5

The Mesozoic.

This time-scale shows the various chronostratigraphical subdivisions of the Mesozoic including system (periods), epochs (including traditional European subdivisions) and stages, together with their upper age limits marked in millions of years (far left).

The Cretaceous (red) is the longest of the three periods. It records the diversification of flowering plants, modern teleost fish and the final extinction of the dinosaurs.

The Jurassic (yellow) is marked by the great radiation of many typical Mesozoic faunal elements including ammonites, dinosaurs, pterosaurs and marine reptiles. The earliest known birds appear in the Late Jurassic.

Many modern marine invertebrate groups greatly diversified in the Triassic (blue) and the first dinosaurs, pterosaurs and ichthyosaurs started to appear. Many archaic groups such as therapsids and temnospondyl amphibians went into decline by the end of the Triassic.

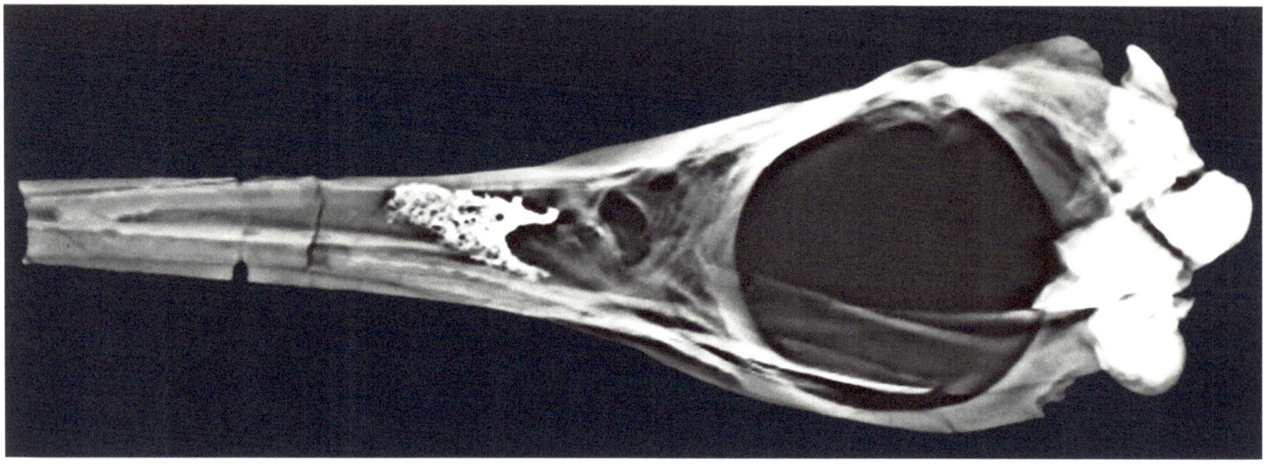

Ichthyosaurs are one of the most characteristic Mesozoic marine reptile groups. This computerised topographic (CT) scan of a juvenile ichthyosaur (*Platypterygius australis*) skull from the Lower Cretaceous of Queensland shows dense calcite seams filling spaces within the wafer-thin internal nasal bones, highlighted in white. Scan image courtesy Royal Adelaide Hospital.

During the Late Jurassic, early teleost fish (modern ray-finned fish) began to diversify, as did some modern cartilaginous fishes including chimaerids (rat-fish) and many advanced sharks.

The final phase of the Mesozoic, the Cretaceous (145.5–65.5 million years ago), was characterised by the development of modern-looking ecosystems. During the earliest part of the period, some familiar Jurassic elements persisted. For example, Early Cretaceous floras were still dominated by conifers, cycads, ferns and ginkgos. By the mid Cretaceous a gradual global cooling trend facilitated the first

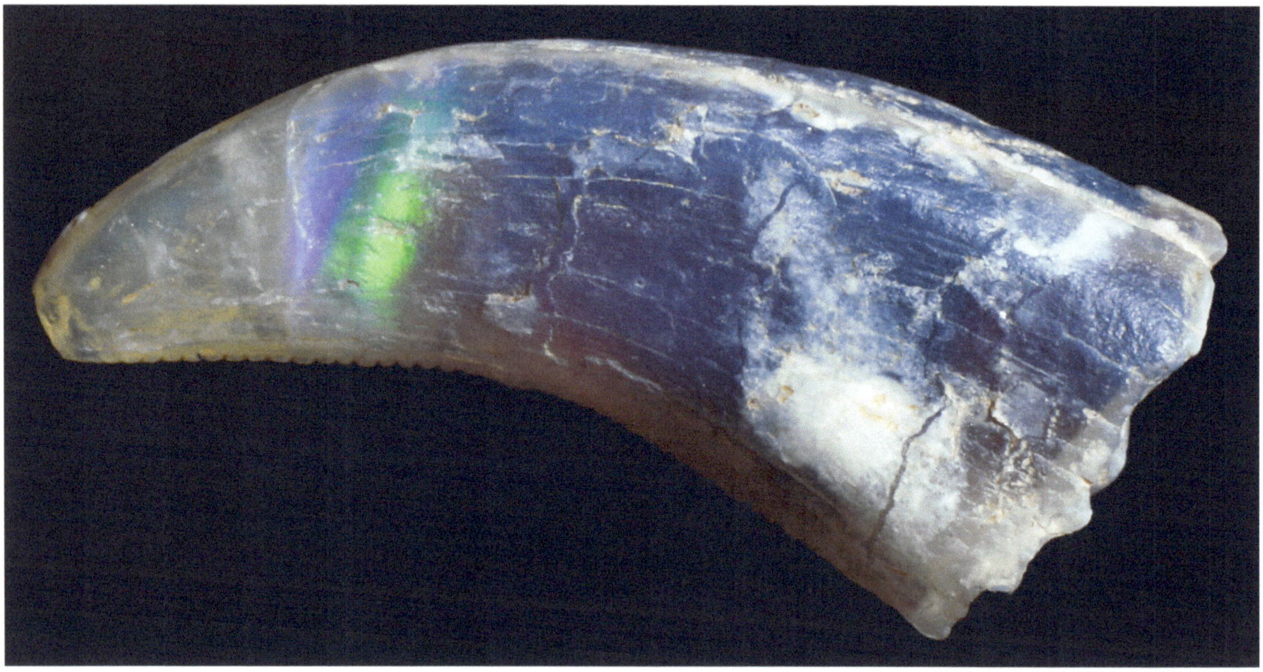

The remains of Australian carnivorous dinosaurs are generally very fragmentary, but the teeth of small theropods are quite common in some deposits and suggest the presence of several distinct groups. This opalised specimen is from Lightning Ridge in New South Wales.

The coiled shells of ammonites are one of the most characteristic Mesozoic fossils. Like the dinosaurs, this group of squid-like animals disappeared at the end of the Cretaceous. This specimen (*Myloceras*) was found near Oodnadatta in northern South Australia.

appearance of angiosperms or flowering plants. These radiated rapidly from their apparent origin in western Gondwana (around eastern Africa or northern South America) and became a major part of global floras by the end of the Cretaceous. The Early Cretaceous also saw the first appearance of crown mammals, the group that includes monotremes (egg-laying mammals, e.g. platypus and echidna), marsupials (pouched mammals, e.g. kangaroos) and placentals (live-bearing mammals, e.g. primates). Monotremes probably originated in Australia, although the first marsupials were found in Asia, and radiated into Gondwana before the end of the Cretaceous. The earliest placentals are traditionally thought to have been Asian in origin; however, the recent discovery of controversial remains in Victoria might indicate that placental mammals also radiated in the southern hemisphere.

Dinosaur faunas during the Cretaceous were very different from those of the Jurassic. Ornithopods, ankylosaurs (armoured dinosaurs) and ceratopsians (horned dinosaurs) became prolific. Sauropods were mainly restricted to the Gondwanan continents. Theropods diversified in the Early Cretaceous, giving rise to bird-like maniraptorans, the group that includes the famous *Tyrannosaurus* and *Velociraptor*. True birds became increasingly common and produced a range of specialised flightless forms adapted to marine diving and ground-dwelling lifestyles.

In the oceans, Cretaceous marine vertebrate faunas included the first true sea turtles and long-snouted polycotylid plesiosaurs – the latter had their earliest representatives in Australia. Ichthyosaurs and giant pliosaurid plesiosaurs were also prolific. However, by the middle Cretaceous these groups had become extinct, possibly in response to competition from other marine reptiles and advanced bony fish. The Late Cretaceous is marked by the explosive radiation of mosasaurs – giant marine lizards that became one of the most successful marine tetrapod lineages of the late Mesozoic. These lived alongside diverse marine invertebrate assemblages, although several extinction events, particularly among cephalopods, mark faunal crises during the middle Cretaceous and the end of the Mesozoic.

Changing Mesozoic geography and climates

Climates during the Mesozoic were more equitable than they are today. Temperatures were generally warmer at high latitudes and seasonality was less pronounced. Much of the climatic stability was driven by the changing configuration of the continents. At the beginning of the Mesozoic (Early-Middle Triassic), the massive supercontinent Pangaea covered approximately one-third of the globe. This continuous landsurface restricted ocean current circulation and prompted the development of drier conditions, especially within inland regions. At about this time, Australia formed a high-latitude peninsular landmass at the south-eastern end of Pangaea, a position it maintained throughout the remainder of the Triassic and Early Jurassic. By the Middle Jurassic an expansive seaway called the Tethys began to widen, separating the southern Gondwanan landmass from the northern Laurasian continents. India started to break away from Australia–Antarctica and, in the northern hemisphere, spreading of the newborn Atlantic Ocean prompted a north–south partitioning of Laurasia. This activity created some barrier seaways but the continents remained in close proximity, allowing the development of homogeneous terrestrial faunas across the globe.

During the Late Jurassic, global climates underwent a moist cooling trend, which continued into the Early Cretaceous. Continental separation and the evolution of distinctive terrestrial faunas became more obvious. The Gondwanan continents had moved apart, with India forming an island landmass (moving north towards Asia) and Australia–Antarctica splitting away from Africa and South America. However, some commonality in faunal elements was still evident between the Gondwanan continents, suggesting that active species exchange may have persisted across the Antarctic Peninsula well into the early Palaeogene.

In the Early Cretaceous, Australia still occupied its southern high-latitude position. Sagging of the continental interior facilitated the incursion of a shallow inland sea and split the exposed land areas into a series of islands. This coincided with a period of isolation in marine and terrestrial assemblages and with marked climate change, during which distinctly seasonal conditions produced montane glacier formation and winter ice build-up along coastal margins. By the beginning of the Late Cretaceous this cold-snap had waned and warm temperatures had returned. Africa separated completely from other Gondwanan continents in the Late Cretaceous, and the islands of New Zealand sheared off from the Australian–Antarctic landmass. The Late Cretaceous was also the time when the inland sea drained off the Australian continent, leaving a series of lowland floodplains and rivers. A narrow seaway subsequently developed along the southern continental margin, flooding what was previously an extensive Early Cretaceous rift valley system. It foreshadowed the eventual separation of Australia from Antarctica, in the middle Palaeogene.

The latest Cretaceous was characterised by another global cooling event. Further continental separation occurred, particularly in the northern hemisphere where the Laurasian landmasses became isolated by major marine incursions. These split the eastern and western halves of North America, and separated eastern North America from Europe–Asia. Geographic segregation led to the development of endemism (restriction to certain areas) in terrestrial faunas. Marine assemblages were more cosmopolitan, with similar groups and

Australia's geographic placement at stages through the Mesozoic.

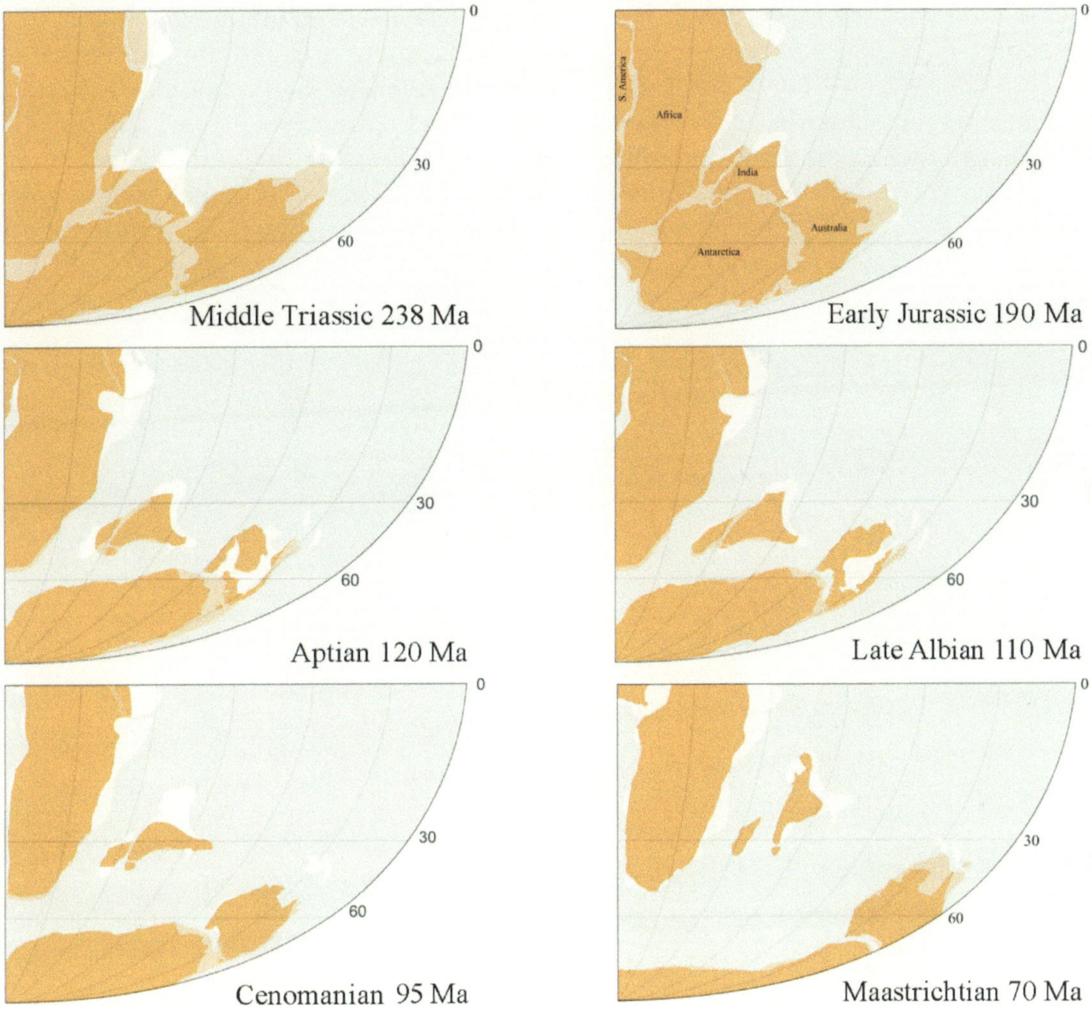

Oceanic areas including continental slope and deep abyssal regions.

Current continent outlines or present landsurface.

Ancient continental landsurface areas through time.

Areas of epicontinental sea inundating current landsurface.

even species occurring at both high and low latitudes. In Australia the latest Cretaceous marine faunas were distinctly transitional, sharing elements with both the northern Tethyan and southern high-latitude regions.

Australian Mesozoic rocks: where and why?

Mesozoic sedimentary rocks are widely distributed in Australia. Their surface exposure or outcrop is dictated by a variety of factors including crustal uplift and weathering. The development of sedimentary basins and the incursion of seas onto the continent have also had major effects on the deposition of Mesozoic sediments.

Australian Triassic surface deposits are predominantly terrestrial in origin. The most extensive were laid down by river and lake systems that flowed across floodplains bordering the eastern (near Sydney) and north-western (west of Broome, Western Australia) margins of the continent in the Early-Middle Triassic. These sediments were derived from areas of Palaeozoic uplift in central New South

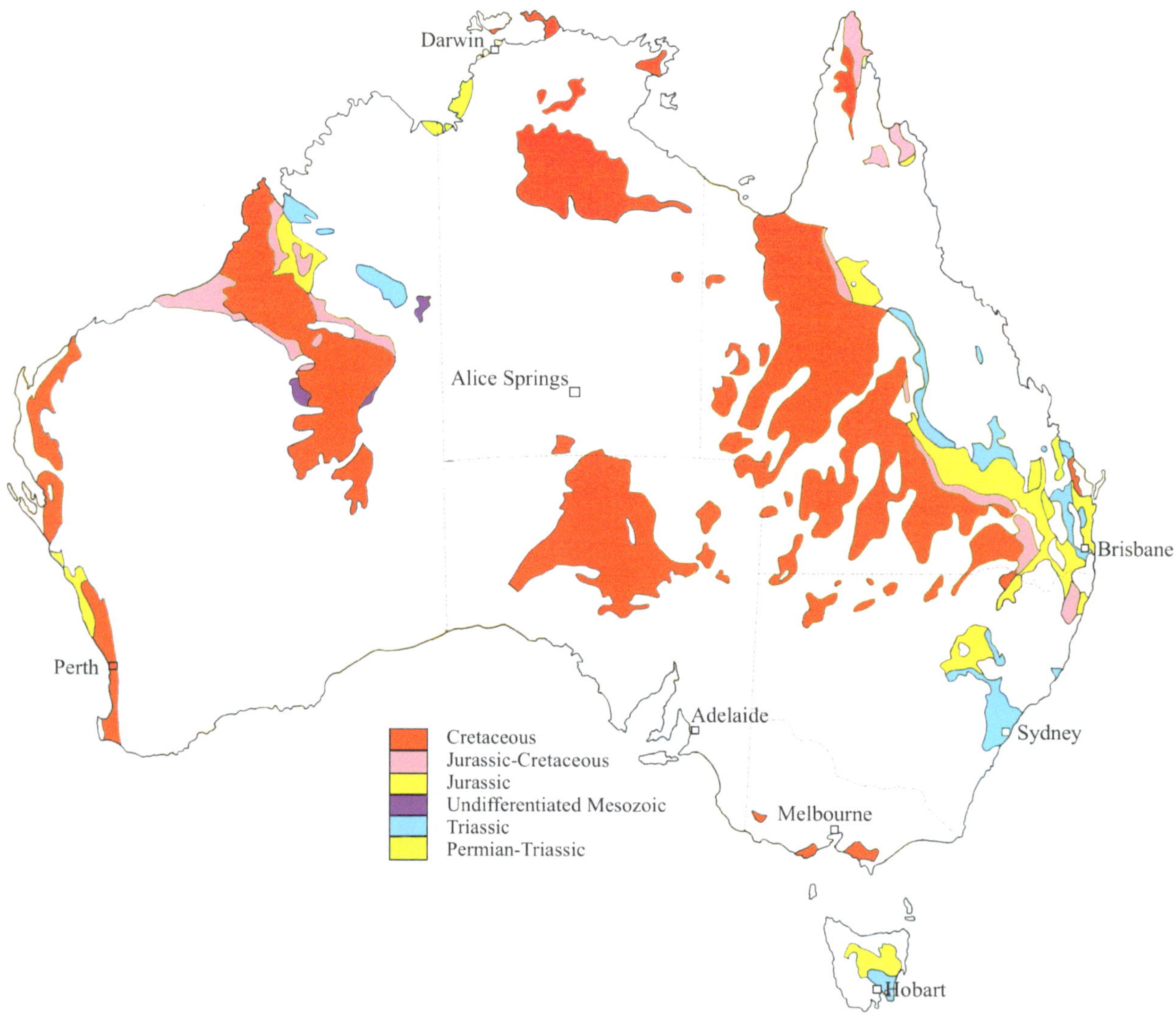

The distribution of Mesozoic rocks within Australia. Most of the exposed outcrop is of Cretaceous age and represents primarily shallow epicontinental marine depositional environments. Triassic and Jurassic continental sediments are also present but are distributed much more sporadically.

Wales and the Kimberley region of Western Australia respectively. Downwarping in south-east Tasmania and Victoria during the early Triassic led to the establishment of lakes and swamps, that in some areas formed coal. Coal also deposited in inland peaty swamps around Leigh Creek in South Australia during the Late Triassic.

Exposures of Australian Triassic marine rocks are uncommon although some occur in north-western Western Australia and in eastern Queensland. These are derived from brackish coastal embayments that invaded areas of late Palaeozoic (Permian) subsidence during the Early-Middle Triassic.

Australian Jurassic rocks are largely derived from terrestrial sources, laid down by lake and swamp systems drawing sediment from the continued erosion of Palaeozoic mountain ranges in eastern Australia. Marine transgressions in north-western Queensland extended close to the present coastline and mark the genesis of the modern Gulf of Carpentaria. Substantial marine inundations persisted along the Western Australian coast, and by the Middle Jurassic they covered much of the present continental margin. Scattered rock outcrops, many of which are overlain by younger sediments, record these environments.

The bulk of Australia's Mesozoic rocks are Cretaceous in age. Major marine transgressions in the earliest Cretaceous inundated the Queensland and northern Western Australian coasts. A shallow inland sea covered much of the Northern Territory and central Queensland. It was not until the middle Cretaceous that the most extensive marine invasion occurred, flooding the Australian continent from the south, west and north and effectively cutting the landsurface into a series of islands. This new inland seaway deposited huge amounts of sediment across central Queensland, northern New South Wales, central South Australia and north-western Western Australia. The south-eastern continental margin remained landlocked, although a rift valley system had developed in response to the initial separation of Australia from Antarctica. This allowed for deposition of sediments by inland rivers which, together with isolated subsurface deposits in northern New South Wales, represent one of the few sources of Australian Lower Cretaceous non-marine fossils.

By the beginning of the Late Cretaceous the inland sea had retreated, leaving a series of low-lying lakes and swamps. Areas of marine inundation persisted across far northern Australia and along the Western Australian coast. Those shallow seas lasted well into the latest part of the Cretaceous and have left extensive deposits of highly fossiliferous calcareous rocks in far north-western Western Australia.

3

DAWN OF AN ERA: LIFE FROM THE TRIASSIC

At the dawn of the Mesozoic 251 million years ago, Australia lay at extreme high latitude and was orientated at right angles to its present geographic position. The southern polar circle was situated in present-day New South Wales and the Western Australian coastline lay close to 60°S palaeolatitude. Triassic polar climates were radically different from those today. The Australian coastal regions experienced generally warm humid conditions; further inland and at lower latitudes, more arid and seasonally monsoonal environments predominated. The landmass itself formed a southern peninsular extremity of the Pangaean supercontinent, and it contacted other Gondwanan continental fragments including India, New Zealand and the western coastal terrain of New Caledonia. New Guinea represented an eastern coastal margin and offshore shelf. Almost all of the current continental landsurface was exposed as dry land, although in the Early Triassic (Induan-Olenekian, ~249 million years ago) shallow marine embayments briefly invaded low-lying coastal areas in the Maryborough Basin of Queensland and the Canning and Perth basins of Western Australia. Throughout the Triassic, Australia's north-western coastal margin was bordered by the vast Meso-Tethys Ocean. In the east, extensive crustal folding and uplift had created a series of uplands running from modern Sydney into south-eastern Queensland. These provided the primary sediment source for south-easterly flowing river systems, depositing massive volumes of sand and mud across the coastal lowland floodplains of the Sydney Basin in the Early and Middle Triassic. Evidence for Early Triassic coastal deltas also exists in the Canning Basin of Western Australia; the deltas were fed by upland areas in the north-western Kimberleys. Other mountainous regions were present in volcanically active south-eastern Queensland (associated with ash deposits and mudflows), central Australia and northern Tasmania where upland areas bordered extensive lowland lake systems and peat swamps.

Hot, dry climates became prevalent in the Late Triassic (Carnian-Norian, 228.7–203.6 million years ago). Nevertheless, subsidence and faulting in upland areas of the Flinders Ranges in South Australia maintained lake complexes and coal-forming peat swamps, suggesting localised humid environments. Similar conditions seem to have existed within montane depressions in the Ipswich Basin of south-eastern Queensland. These are now a lucrative source of coal.

Most Australian Triassic fossil-bearing strata are of Induan (Lower Triassic) to lower Landinian (Middle Triassic) age. Significant Induan-Olenekian deposits include the fluvial Arcadia Formation of south-eastern Queensland. This unit crops out in the famous Crater locality near Rolleston in the Bowen Basin and has yielded many important amphibian and reptile fossils. Other rich Induan-Olenekian units include the Blina Shale and contemporary Kockatea Shale. These are exposed along the coast and on the offshore continental

The Triassic 'red beds' of the Arcadia Formation at Duckworth Creek near Rolleston in south-central Queensland. These deposits were laid down by rivers in a seasonally monsoonal environment and have been a productive source of vertebrate fossils, particularly temnospondyl amphibians. Image courtesy Anne Warren.

shelf of Western Australia and have produced the remains of fossil invertebrates, amphibians and plants that inhabited a brackish delta and near-shore marine embayments.

A wealth of Olenekian to lower Landinian (Lower to Middle Triassic) fossils have been derived from the extensive fluvial and lacustrine sediments of the Narrabeen Group, Hawkesbury Sandstone and Wianamatta Group in the Sydney Basin of central-eastern New South Wales. Well-preserved plant, invertebrate and vertebrate specimens (especially fish) have been recovered from these deposits, which represent one of the most continuous records of Triassic organisms known from the Australian continent.

Carnian and Norian (Upper Triassic) fossils have been recorded from coal-bearing strata of the Leigh Creek Coal Measures (most extensive in the Telford Basin) in northern South Australia. This upland region was dominated by peat swamps and lake systems 235 million years ago. It preserves plants, invertebrates (mainly bivalves) and rare vertebrate remains, recovered from tailings heaps left by coalmines.

Significant Australian Triassic deposits and localities

Among the oldest of the Australian Triassic units is the Arcadia Formation (Rewan Group), a series of terrestrial sediments that record the transition from moist swampy coal-forming conditions in the latest Permian to more arid highly seasonal climates characteristic of the inland Early Triassic.

The Arcadia Formation comprises up to 500 m of predominantly red-coloured shales (red beds) that are interbedded with cross-bedded sandstones

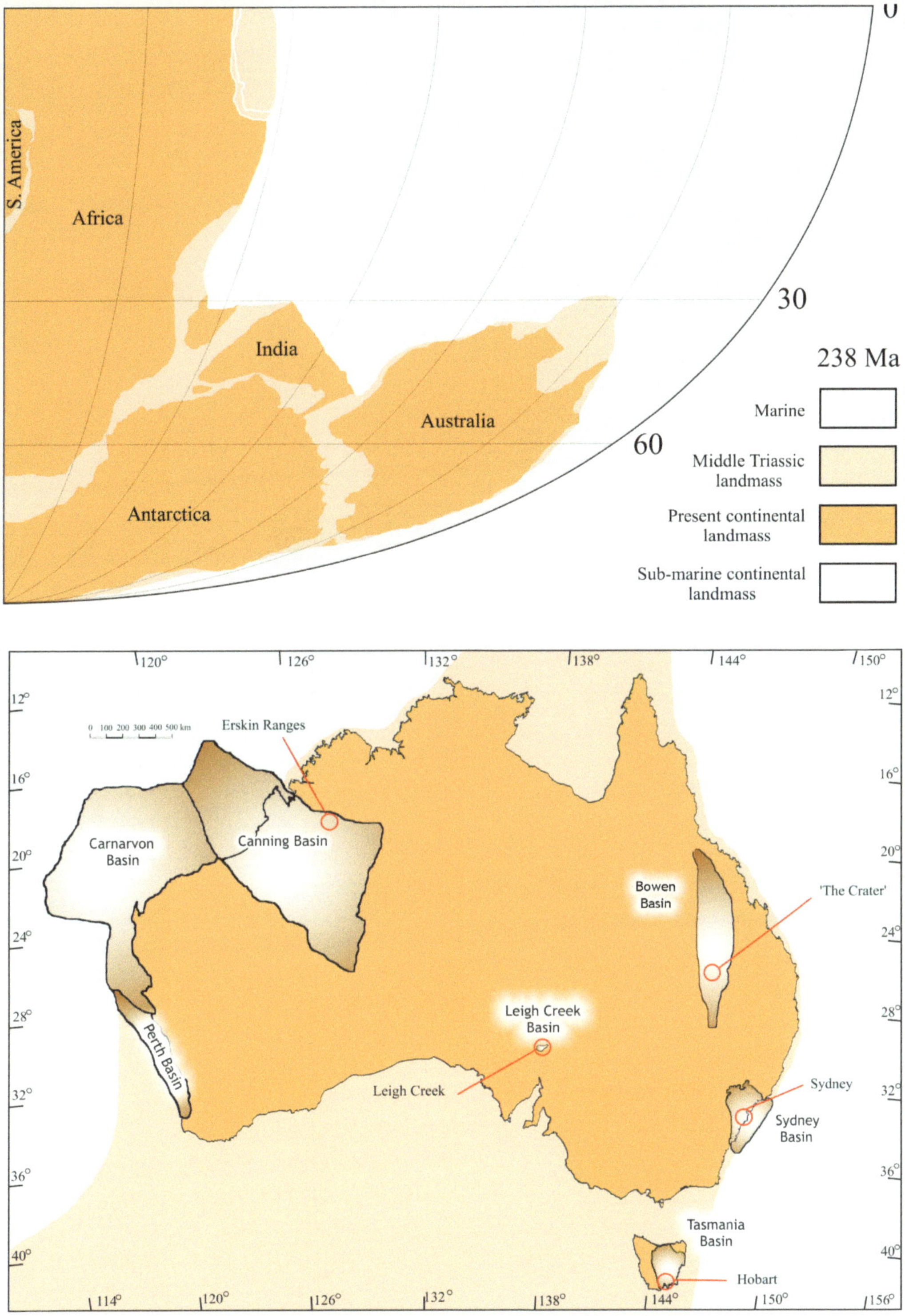

(Top) The geographic position of the Australian continent during the Middle Triassic Anisian stage, ~238 million years ago. (Bottom) Principal Triassic fossil localities (indicated in red) occur mainly throughout the depositional basins of eastern Australia and in the Canning Basin of far north-western Western Australia.

The Crater is arguably one of Australia's most famous Triassic fossil localities. Its red beds, which form part of the Arcadia Formation outcrops near Rolleston in south-central Queensland, have yielded a large number of plant, fish, amphibian and reptile fossils. Image courtesy Anne Warren.

(denoting stream channels) that crop out within the Bowen Basin of south-eastern Queensland. Deposition of the Arcadia Formation occurred in meandering streams, which flowed across vegetated floodplains. The red coloration of the sediments indicates iron-rich oxidising conditions and a semi-arid palaeoclimate, probably with marked seasonal temperature fluctuations and summer monsoons. The Bowen Basin lay at high palaeolatitudes (over 80°S) during the earliest Triassic and was located well inland, several hundred kilometres from the coast. A rich fossil record of spores and pollen, with macroflora and vertebrates, indicates a predominantly Induan-Olenekian age.

The Blina Shale also corresponds to an Early Triassic riverine setting, probably part of an extensive brackish-marine delta system that spread across the north-western Canning Basin of Western Australia during the Late Permian to Early Triassic. The Blina Shale is composed of predominantly yellow shales and claystones, exposed as low hills and bluffs in the Erskine Ranges. The Blina Shale is notable for producing vertebrate bone beds up to 3 m thick. These probably accummulated as current velocity declined in backwaters and coastal lagoons. An earliest Triassic (Induan-Olenekian) age has been suggested for the Blina Shale on the basis of its plant and microfossil assemblages.

The Kockatea Shale characterises the Lower Triassic marine rocks deposited along the Western Australian coast. This sequence of shallow near-shore marine strata crops out in limited areas around the Greenough River near the township of Geraldton (Perth Basin) in south-western Western Australia. The Kockatea Shale is also encountered in offshore boreholes and is thought to have

The sea-cliffs at Long Reef in the northern suburbs of Sydney. These deposits represent part of the Bulgo Sandstone and have produced a tantalising array of Lower Triassic plant and vertebrate fossils. Some of the most spectacular have been fragmentary remains of the giant (3 m) temnospondyl amphibian *Bulgosuchus*.

accumulated during a short-lived marine incursion onto the western margin of the Australian continent. The Kockatea Shale is considered a stratigraphical contemporary of the deltaic Blina Shale. It has produced cephalopod and bivalve fossils within greyish shale that grades upwards from fine basal sandstones. Mollusc faunas, particularly ammonoids, have been used to demonstrate a probable Induan to Olenekian age for the unit.

The Australian Early to Middle Triassic terrestrial record is best typified by the Narrabeen Group, Hawkesbury Sandstone and Wianamatta Group of central-eastern New South Wales. These units constitute a continuous series of shales, sandstones and claystones that are widely distributed throughout the Sydney Basin and represent prograding shoreface/coastal plain sedimentation. The predominantly Lower Triassic (Olenekian) Narrabeen Group is the lowest of the three major subdivisions and is conformably overlain by the Middle Triassic (Anisian) Hawkesbury Sandstone (together with the transitional Mittagong Formation) and uppermost Anisian-Landinian Wianamatta Group. The age ranges of these strata have been determined using rich microfossil (spores and pollen) and plant/vertebrate assemblages.

Key fossiliferous units within the lowermost Narrabeen Group include the Olenekian Bulgo Sandstone at Long Reef in the northern suburbs of Sydney, and the Lower to Middle Triassic (upper Olenekian-lower Anisian) Terrigal Formation near Gosford, north of Sydney. These rocks are exposed along extensive coastal sea-cliffs and adjacent rock platforms. The Bulgo Sandstone and Terrigal Formation are composed of fine-grained shales and sandstones deposited by a south-easterly flowing stream system on a coastal floodplain. The Anisian Hawkesbury Sandstone is another conspicuous rock unit in the Sydney region, variously exposed as sea-cliffs and in quarries throughout suburban areas. A fluvial environment has been envisaged for the unit, which is typified by interbedded coarse sandstone and minor mudstone facies. These are an abundant source of microfossils (spores and pollen), plant debris and invertebrate/vertebrate remains. The Hawkesbury Sandstone is thought to reflect peak flood and low-energy washout events within a massive river delta that emptied off the eastern margin of the Australian landmass during the Middle Triassic.

One of the most fossiliferous sequences within the Wianamatta Group is the middle Anisian-Landinian Ashfield Shale at the famous St Peters brick pits in the inner western suburbs of Sydney. This deposit, which was mined extensively in the early 20th century, has yielded many important invertebrate and vertebrate fossils from its carbonaceous shale and claystone layers. The Ashfield Shale is regarded as a freshwater lacustrine palaeoenvironment that was steadily inundated by brackish and shallow marine conditions over time.

Australian Late Triassic fossils are recorded from several units, most notably the Leigh Creek Coal Measures near Leigh Creek in the Flinders Ranges of north-eastern South Australia. These deposits comprise ~1000 m of interbedded coal seams (each 10–20 m thick) and laminated siltstones laid down in five Triassic intramontane basins, the largest of which – the Telford Basin – developed in response to localised subsidence and faulting. The depositional settings were predominantly lacustrine with peat swamps and episodic influx of fluvial systems during later stages of basin infill. The Leigh Creek Coal Measures are considered Carnian to Norian in age on the basis of plant macrofossil and spore/pollen fossil assemblages.

Land plants

Vegetation during the Triassic was distinctly provincial. Southern Gondwanan and northern Laurasian floras were differentiated by climatic factors including monsoons and extreme seasonality, caused by symmetrical placement of Pangaea over the equator. In Australia, the Triassic plant macrofossil record is based on particularly well-documented assemblages from the Lower to Middle Triassic Sydney Basin and Late Triassic coal measures of the Telford Basin in South Australia and Ipswich Basin in south-eastern Queensland. The general composition of these floras closely resembles contemporary communities elsewhere in Gondwana, with the dominant low-growing seed-fern *Dicrodium* occurring in all assemblages from lowland heath to dry upland woodlands. Other common Australian Triassic plants include voltziacean and primitive podocarpaceous conifers, together with cycadophytes (cycads and relatives), ginkgoaleans (ginkgos), equisetaleans (horsetails) and lycophytes (clubmosses) as understorey elements.

The earliest Australian Triassic macrofloras occur in the Olenekian-Landinian Sydney Basin. They comprise a diverse range of arborescent lycophytes (*Pleuromeia*), seed-ferns (*Dicrodium*) and equisetaleans as well as rare voltziacean (*Voltziopsis*) and podocarp (*Podocarpus*) conifers. This vegetation occupied coastal wetland environments and has been interpreted as an indicator of warm humid climatic conditions. Indeed, key elements such as *Pleuromeia* are thought of as mangrove-like plants, growing in partly submerged monodominant thickets along watercourses and waterlogged swampy areas. Adjacent lake margins were probably surrounded by *Dicrodium*-dominated heath zones with conifers restricted to well-drained upland areas. Correlations based on macrofossils, spores and pollen show that similar floras (e.g. *Dicrodium* and *Pleuromeia*) existed in Western Australia and subarid inland areas of eastern Queensland. However, some variability is evident: inland vegetative types

Remains of the seed-fern *Dicrodium* represent some of the most characteristic Gondwanan Triassic fossils. Foliage of this treefern-like plant has been found throughout Australia, suggesting that *Dicrodium* was probably a common element in many Australian Triassic floras. This frond (with extracted imprint) is from the Lower Triassic shales of Brookvale, near Sydney.

(exemplified by material from the Arcadia Formation and Blina Shale) reflected a more seasonal flora mainly restricted to established watercourses and influenced by summer monsoonal rainfall.

During the Middle Triassic, uniform arid-zone floras developed across the Australian continent. *Dicrodium* species diversified into arid-adapted forms (e.g. *Dicrodium tripinnata*) with thickened leaf cuticles and reduced spiny leaves to reduce water loss in hot dry climatic conditions. Contemporary macrofloras from coastal river deposits in the Hawkesbury Sandstone, particularly the Beacon Hill Quarry at Brookvale near Sydney, preserve evidence of understorey foliage comprising seed-ferns (*Dicrodium*), ferns (*Rienitsia spatulata*), equisetaleans (*Neocalamites hoerenis*), ginkgophytes and cycadophytes. Podocarp conifers (*Rissikia*, which persisted into the Jurassic) formed the principal canopy elements.

Late Triassic macrofloras were broadly similar across Gondwana with seed-ferns, locally abundant volzialean conifers, ginkgophytes, pleuromeian lycophytes and ferns predominating. These vegetative types have also been recovered from Australian Carnian-Norian coal-forming swamp deposits in the Telford (Leigh Creek Coal Measures) and Ipswich basins (Brassall Subgroup, Ipswich Coal Measures) with evidence of seed-ferns (*Dicrodium*), osmundacean ferns, ginkgophytes (*Ginkgoites*), podocarp conifers (*Rissikia*) and the bennettitalean *Otozamites* (a distinctive Mesozoic cycadophyte).

Marine and non-marine invertebrates

Australian Triassic macroinvertebrate faunas are well-documented. One of the best-known non-marine occurrences comes from the Triassic Ipswich Coal Measures at Mt Crosby in south-eastern Queensland. These are extremely rich in fossil insects including cockroaches (e.g. *Triassoblatta*, which constitutes around 50% of the assemblage and incorporates several species from Triassic deposits throughout eastern Australia), sawflies (e.g. *Archexyela crosbyi*), scorpionflies (e.g *Mesochorista*, dominant in the Sydney Basin assemblages), caddisflies (*Cladochoristella bryani*), orthopterans including grasshoppers (*Triassolocusta leptoptera*, also recovered from the Sydney Basin), dragonflies (odonatans) and beetles (or coleopterans, known from isolated elytra and wings).

A well-preserved Middle Triassic (Anisian-Landinian) non-marine invertebrate fauna is also known from the Hawkesbury Sandstone at Beacon Hill, and from the Ashfield Shale (Wianamatta Group) at St Peters in suburban Sydney. These have produced several types of unionoid bivalve (*Protovirgus dunstani*,

Foliage of various Triassic plants. (Clockwise from top left) The seed-fern *Dicrodium superbum*, the drought-adapted species *Dicrodium feistantelli*, a frond of the bennettitalean *Otozamites*, the podocarp conifer *Rissikia*, the ginkgophyte *Ginkgoites bidens* and the osmundacean fern *Cladophlebis australis*. All specimens are from the Middle Triassic Esk Formation of Queensland.

Protovirgus brookvalensis and *Unionella wianamattensis*), together with crustaceans (the isopod *Protamphisopus winamattaensis* and a conchostracan) and a range of insects; some, such as the giant titanopteran *Clatrotitan*, had a wingspan of ~300 mm. A few freshwater bivalves (*Prohyria eyrensis* and *Megalovirgus jaenschi*) and unidentified insect wing impressions have been identified from the Carnian-Norian Leigh Creek Coal Measures in South Australia. Contemporaneous deposits in the Ipswich Coal Measures of south-

Shells of Late Triassic freshwater unionoid bivalves. (Left and middle) *Prohyria eyrensis* from the Carnian-Norian Leigh Creek Coal Measures of north-eastern South Australia. (Right) *Protovirgus brookvalensis* from the Anisian-Landinian Hawkesbury Sandstone of Sydney, New South Wales.

eastern Queensland have yielded specimens of the unionoid bivalve *Mesohydridella ipsviciensis.*

The current record of Australian Triassic marine macroinvertebrates is very limited. Only four isolated assemblages have been recorded from offshore boreholes in the Lower Triassic Kockatea Shale of the Perth Basin and undifferentiated Middle Triassic sediments of the Sahul Shoals off the Western Australian coast. Onshore outcrops of the Kockatea Shale in the nearby Mt Minchin area (Perth Basin) and the Broweena Formation near Maryborough (Maryborough Basin)/ Traverston Formation near Woodnum (Gympie Basin) of south-eastern Queensland have also yielded some remains. The Kockatea Shale assemblage includes ammonoid (*Anasibrites*, *Arctoceras*, *Glyptophiceras*, *Gyronites*, *Hemiprionites*, *Ophiceras*, *Proptychites*) and bivalve (*Nuculana*) genera common to the southern Tethyan shelf (e.g. Afganistan, Pakistan, Himalayas and Timor) and Pacific regions (Japan, western North America) of Gondwana. Invertebrates from the Broweena and Traverston formations correlate with the upper Kockatea Shale. They comprise southern Tethyan/ Pacific ammonoids (with endemic species *Anaflemingites armstrongi*, *Dieneroceras woondumense*, *Latisageceras woondumense*, *Paranorites queenslandicus*), together with bivalves (*Bakevellia*, *Nuculanella*), indeterminate gastropods (snails) and ophiuroids (brittle-stars). The Middle Triassic Sahul Shoals have produced ammonoid remains tentatively attributable to *Nicomedites*, a genus reported elsewhere from Asia Minor and perhaps suggesting a lower Anisian age for the deposit.

Non-marine fish

Australian Triassic fish fossils are known only from non-marine deposits. However, the documented faunas are extremely rich and include over 30 genera and species. The earliest Triassic (Induan-Olenekian) assemblages have been recorded from the Arcadia Formation of Queensland, fluvial/ lacustrine Knocklofty Formation (Tasmania Basin) near Hobart in Tasmania, and the Blina Shale in Western Australia. These units have produced fossils of the globally distributed *Saurichthys* (a long-snouted eel-like predator >1 m long), the palaeoniscoid (primitive ray-finned fish) *Acrolepis* and

Middle Triassic insects. (Top) The giant dragonfly-like titanopteran *Clatrotitan scullyi* from the Hawkesbury Sandstone of Brookvale, New South Wales, showing a wing nearly 150 mm long. (Bottom) *Eoscartoides bryani* from the Esk Formation, Queensland.

numerous lungfish toothplates/skull elements attributable to several taxa including *Aphelodus anapes*, *Namatozodia piticanta* and the geographically widespread *Ptychoceratodus philippsi*.

The best-known Australian Triassic fish material comes from the Sydney Basin. The Early to Middle Triassic (Olenekian-lower Anisian) Terrigal Formation near Gosford, north of Sydney, has yielded numerous specimens including sharks, lungfish and actinopterygians (ray-finned fish). These are often complete and compressed onto a single rock layer. Described taxa incorporate the cosmopolitan freshwater shark *Xenacanthus* and deep-bodied lungfish *Gosfordia truncata*, which also ranges up into the Middle Triassic Hawkesbury Sandstone. Much of the remaining Terrigal Formation fish fauna is dominated by primitive actinopterygians such as:

- the saurichthyid *Saurichys gracilis*;
- endemic palaeoniscoids *Apatolepis australis* and *Myriolepis latus*;
- several species of *Palaeoniscum*, a taxon known elsewhere from Palaeozoic-Triassic deposits in Europe, North America and Asia;
- the perleidiform *Cleithrolepis granulata*;
- the widespread semionotid *Semionotus*.

Some of these taxa (e.g. *Cleithrolepis*, *Myriolepis*, *Palaeoniscum*) also occur in the Middle Triassic (Anisian-Landinian) Hawkesbury Sandstone and Ashfield Shale (Wianamatta Group). The famous Brookvale quarry near Sydney and St Peters brick pits in the inner western suburbs of Sydney have also yielded many important forms, including:

Body of the lungfish *Gosfordia truncata*, clearly showing the shape of the tail fin. This specimen is from the Somersby Falls area in New South Wales.

- a diverse range of palaeoniscoids – *Agecephalichthys, Belichthys, Leptogenichthys, Megapteriscus* and *Mesembroniscus*;
- the redfieldiiforms *Brookvalia gracilis* (a member of the endemic Brookvaliidae) and *Geitonichthys ornatus* (a redfieldiid also known from South Africa and North America);
- the perleidiforms *Manlietta crassa* and *Procheirichthys ferox*;
- advanced actinopterygians such as the semionotiform *Enigmatichthys attenuatus* and parasemionotiform *Promecosomina formosa*.

The lungfish *Gosfordia truncata* and *Ariguna formosa*, together with the saurichthyid *Saurichthys parvidens*, are also represented by well-preserved remains. Sharks have been recorded only from the Ashfield Shale at St Peters with one nearly complete skeleton (>1.5 m long) assigned to the xenacanthid *Xenacanthus parvidens*.

The only known Late Triassic fish fossil is a tail section showing a distinctive scale patterning, from the Carnian-Norian intramontane swamp deposits of the Leigh Creek Coal Measures, South Australia. This specimen was described as *Leighiscus hillsi* and is of uncertain taxonomic affinities.

Amphibians

Australian Triassic continental faunas were dominated by temnospondyl amphibians. These crocodile-like predators reached over 3 m in length and appear to have occupied mainly aquatic niches in freshwater and estuarine habitats. Temnospondyls reached a peak of diversity during the Permian and Triassic, but declined markedly from the Triassic-Jurassic boundary, finally becoming extinct in the mid-Cretaceous.

Evidence of Australia's earliest Triassic (Induan-Olenekian) temnospondyls come from the Arcadia Formation (Bowen Basin) and laterally equivalent Rewan Formation (Galilee Basin) near Rolleston in southern Queensland. These rocks have yielded a range of taxa including:

- the widespread mastodonsaurid *Watsonisuchus* (represented by at least two species), a large

Fish fossils from the middle Triassic Hawkesbury Sandstone of Sydney. (Top) The disc-shaped *Cleithrolepis granulata*. (Bottom) *Promecosmina formosa*, one of the most common fish in many of the Sydney Basin deposits.

crocodile-like predator known from South Africa, India and Europe;

- the rhytidosteids (a well-known Gondwanan group) *Acerastea wadeae*, *Arcadia myriadens* and *Rewana quadricuneata* and the tiny *Nanolania anatopretia* (skull length <50 mm);
- the primitive brachyopid (a globally distributed family) *Xenobrachyops allos*;

Middle Triassic fish fossils from the Hawkesbury Sandstone of Brookvale near Sydney. (Top) *Saurichthys gracilis*, an eel-like predator found in Triassic deposits throughout the Sydney Basin. (Bottom) *Brookvalia parvisquamata*, a primitive fish that is unique to Australia.

- a lydekkerinid, *Lydekkerina huxleyi*, found in the Rewan Formation and in age-equivalent sediments in South Africa;
- the possible plagiosaur *Plagiobatrachus pustulatus*, an obligate aquatic temnospondyl best known from Africa and Europe;
- the basal chigutisaurid *Keratobrachyops australis*, a family restricted to Australia and South America;
- a tiny insectivorous lapillopsid known only from Australia and India, *Lapillopsis nana*;
- an early trematosaurid, *Tirraturhinus smisseni*. Members of this cosmopolitan group have also been reported from the overlying Glendial Formation and Blina Shale of Western Australia.

These largely endemic species represent some of the most primitive members of their respective

Skull of the primitive brachyopid temnospondyl amphibian *Xenobrachyops allos* from the Lower Triassic Arcadia Formation, near Rolleston in southern Queensland. Specimen is shown in dorsal (left) and ventral (right) views.

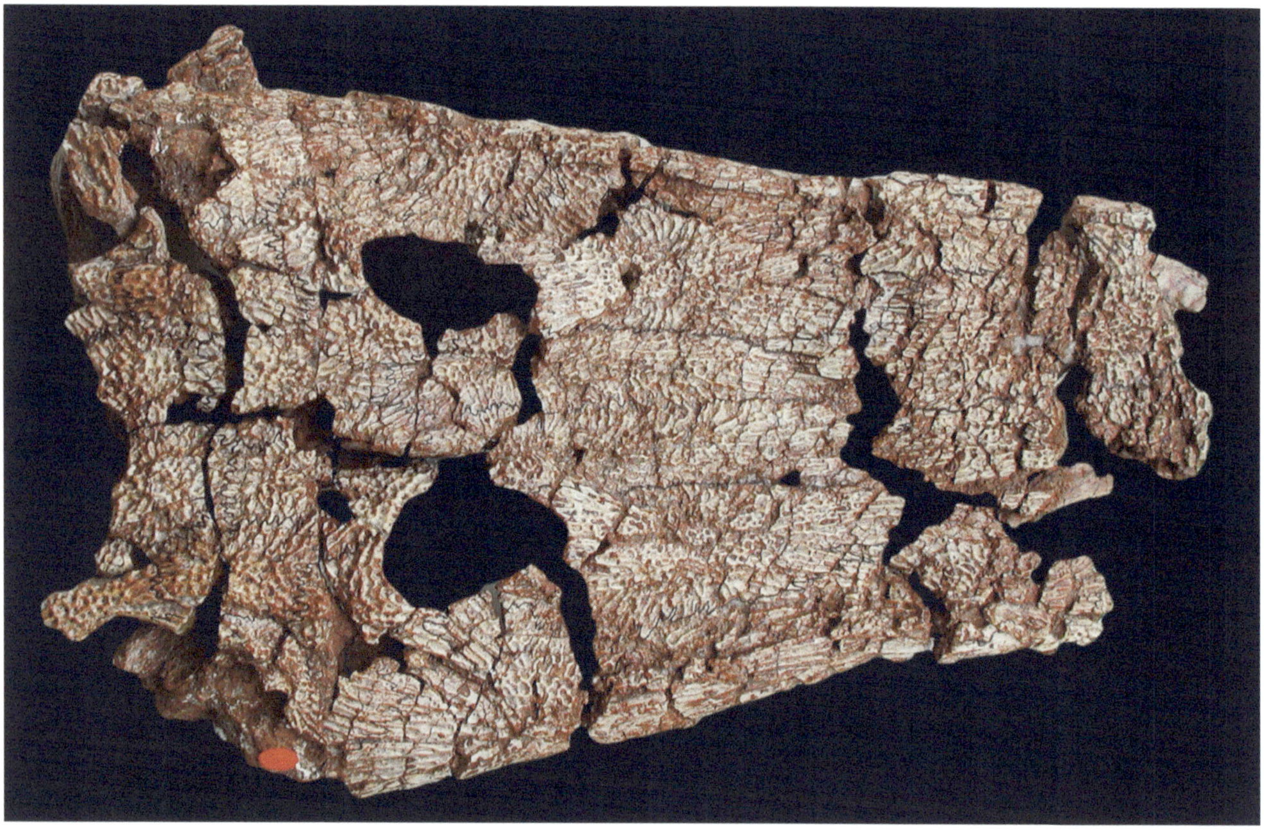

Skull of the mastodonsaurid temnospondyl amphibian *Watsonisuchus* from the Lower Triassic Arcadia Formation, near Rolleston in southern Queensland. Specimen is shown in dorsal view.

Skulls of various Triassic temnospondyls from the Knocklofty Formation near Hobart, Tasmania. (Top) The brachyopid *Banksiops townrowi*. (Middle) The lydekkerinid *Chomobatrachus halei*. (Bottom) The rhytidosteid *Derwentia warrenii*.

family-level groupings and thus imply a probable Gondwanan (or perhaps Australian) origin for many Mesozoic temnospondyl lineages.

Induan-Olenekian amphibian assemblages from the fluvial/lacustrine Knocklofty Formation (Tasmania Basin) of Tasmania and contemporary deltaic Blina Shale (Canning Basin) of Western Australia are dominated by representatives of widespread lineages. However, mastodonsaurids (*Watsonisuchus* found in the Blina Shale) and brachyopids (*Blinasaurus henwoodi* and *Banksiops townrowi*, both flat-skulled suction-feeders on small fishes) are comparatively rare and chigutisaurids are notably absent. The most prolific temnospondyls are small (~1 m long) aquatic fish-eaters, such as the lydekkerinid *Chomobatrachus halei* and the rhytidosteid *Deltasaurus kimberleyensis*, which has been recorded from the Knocklofty Formation, Cluan Formation of Tasmania and Blina Shale. Smaller-bodied rhytidosteids including *Derwentia warrenii* from the Knocklofty Formation (skull length ~90 mm) and *Deltasaurus pustulatus* (skull length ~110 mm), recovered from a drillcore in the marine Kockatea Shale near Geraldton in Western Australia, probably were active predators on small fish and invertebrates. Terrestrial predatory niches appear to have been occupied by a tiny lapillopsid, *Rotaurisaurus contundo* (skull length ~20 mm) from the Knocklofty Formation. Large-bodied temnospondyls included the endemic *Erythrobatrachus noonkanbahensis* (~2 m long), a trematosaurid from the Blina Shale. This long-snouted fish-eater seems to have inhabited coastal estuarine and near-shore marine habitats.

During the Early-Middle Triassic, Australian temnospondyl faunas underwent a significant drop in diversity with only mastodonsaurids and brachyopids present in the coastal river system deposits of the Sydney Basin. One of the oldest described taxa is the giant mastodonsaurid *Bulgosuchus gargantua*, established on the basis of a femur and incomplete mandible from the Olenekian Bulgo Sandstone at Long Reef in the northern suburbs of Sydney. This specimen, with a jaw probably in excess of 1 m long when complete, represents one of the largest Early Triassic temnospondyls yet discovered. The Long Reef locality has produced numerous other bones and teeth of probable mastodonsaurids. Several near-complete skeletons of mastodonsaurids and brachyopids have been recovered from the Olenekian-Anisian Terrigal Formation at Gosford, north of Sydney. The mastodonsaurid remains include skulls attributable to *Watsonisuchus* and an indeterminate large mandible. The brachyopids comprise a skull, pectoral limb girdle and incomplete vertebral column assigned to *Platycepsion wilkinsoni*, and a second (unnamed) taxon represented by a well-preserved skeleton with carbonised body-outline.

Middle Triassic temnospondyl material from the Anisian Hawkesbury Sandstone near Brookvale in Sydney comprises a skull and fragmentary pectoral limb girdle of the mastodonsaurid *Subcyclotosaurus brookvalensis*, indeterminate bone fragments and amphibian footprints. The Anisian-Landinian lake deposits of the Ashfield Shale (Wianamatta Group) have yielded trackways and an isolated skull of the small brachyopid *Notobrachyops picketii*, found in a brick quarry in the southern Sydney suburb of Mortdale. A second indeterminate large brachyopid has been found at St Peters in inner-western Sydney. That locality has also yielded the spectacular 2.25 m long skeleton of *Paracyclotosaurus davidi*, one of the most complete mastodonsaurid specimens ever recovered. Short crocodile-like limbs and a massive flat skull with dorsally facing orbits suggest that *Paracyclotosaurus* was an ambush predator, perhaps lying in wait on the bottom of lakes and catching fish and other temnospondyls in the suction created by rapid opening of its massive jaws.

The only Late Triassic temnospondyl occurrence in Australia is a poorly preserved pectoral limb girdle from the Carnian-Norian intramontane swamp deposits of the Leigh Creek Coal Measures in South Australia. This specimen is non-diagnostic but displays some features in common with chigutisaurids.

Marine reptiles

Triassic marine reptile fossils are almost totally unknown in Australia. A single potential occurrence, based on several bone fragments tentatively attributed to an ichthyosaur, has been reported from the Induan-Olenekian Blina Shale, Western Australia. Ichthyosaurs were a group of superficially fish-like marine reptiles known from Triassic to Cretaceous deposits around the world. Unfortunately, the veracity of the Australian Triassic ichthyosaur material is difficult to assess because the

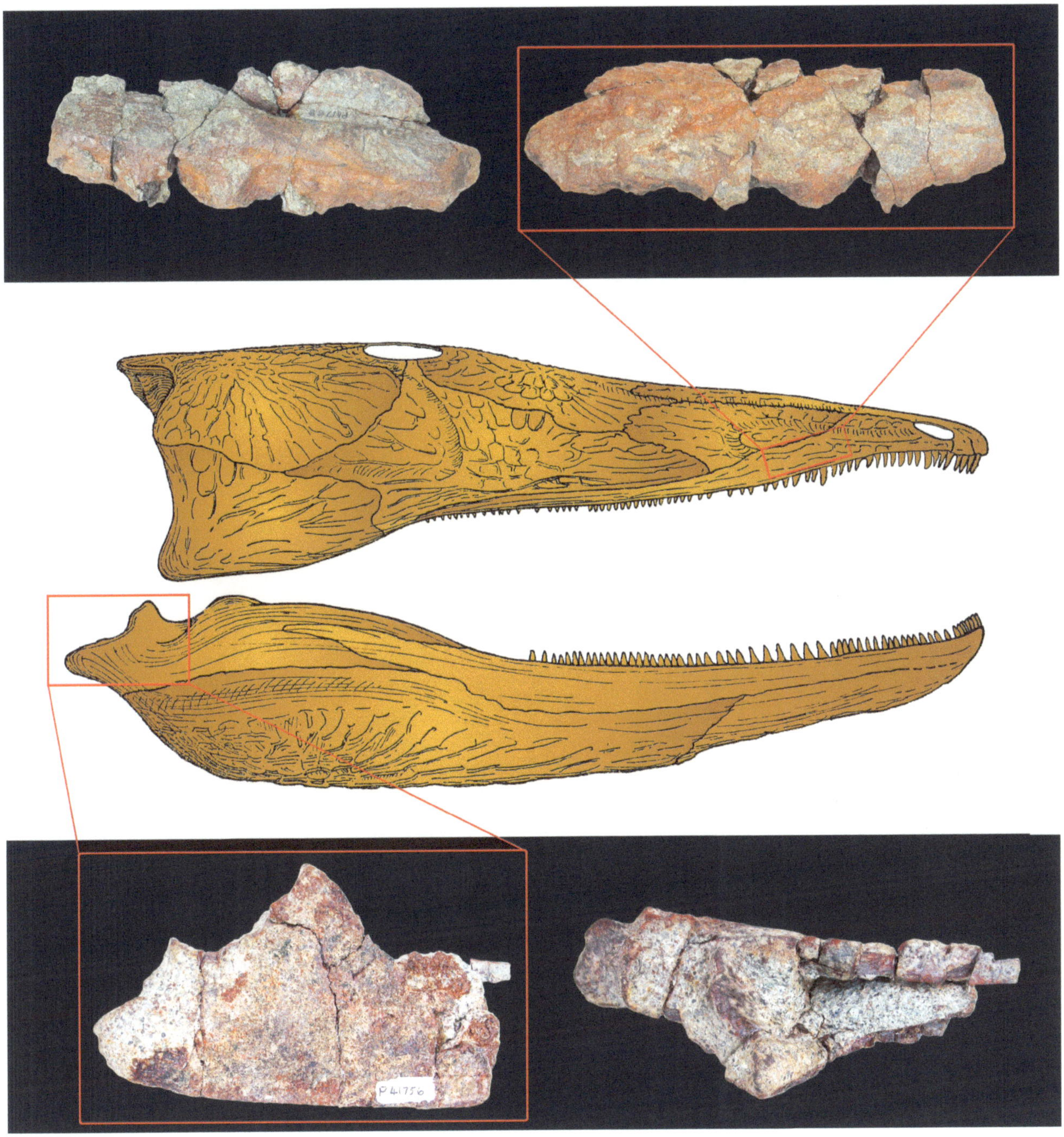

Fragmentary skull bones of mastodonsaurids from the Lower Triassic Bulgo Sandstone, at Long Reef in the northern suburbs of Sydney. (Top) A section of maxilla shown in (left) internal and (right) external views. (Bottom) Rear section of mandible in (left) internal and (right) dorsal views to show the jaw articulation. Skull reconstruction after Watson (1958).

specimens have never been fully described or figured. Consequently, until it can be verified by a more thorough re-examination, the taxonomic assignment of the Blina Shale 'ichthyosaur' remains must be regarded with caution.

Lizard-like reptiles

The Australian Triassic landscape was populated by a variety of lizard-like reptiles. One group, procolophonids, were primitive herbivores with simple peg-like teeth fused directly to the jaw. Procolophonids

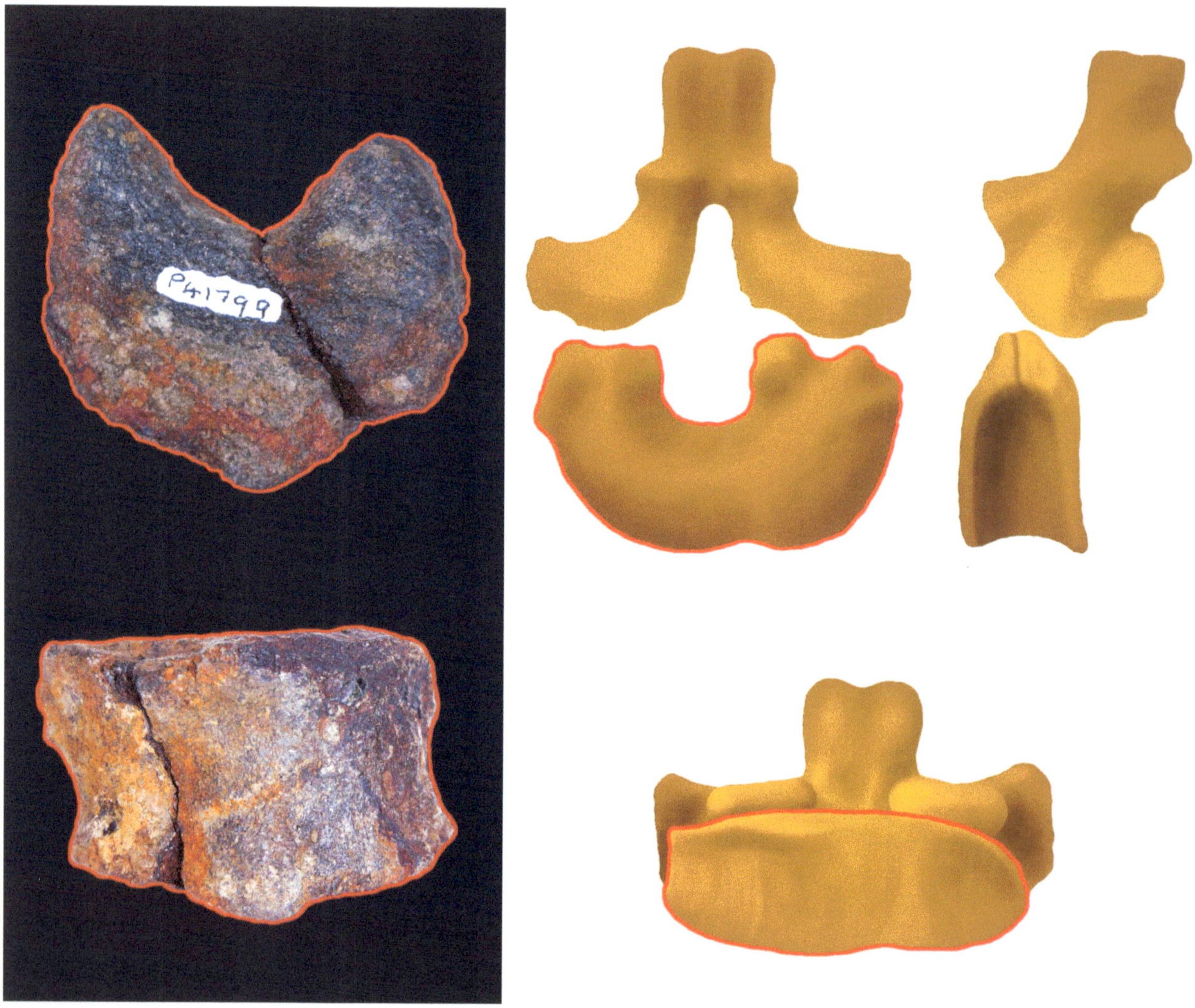

Anterior (top left) and ventral (bottom left) views of a crescent-shaped vertebral centrum typical of temnospondyl amphibians. The reconstruction (after Watson 1958) shows how the centrum and neural arch would have been separated by cartilage in life. The fossil specimen was recovered from the Lower Triassic Bulgo Sandstone at Long Reef in the northern suburbs of Sydney.

had a global distribution throughout much of the Permian and Triassic, and a number of incomplete specimens are known from the Induan-Olenekian Arcadia Formation of southern Queensland. The Australian procolophonid material has not been documented in detail but shows some similarities to taxa from Africa, South America and Antarctica.

Kadimakara australiensis is a small lizard-like reptile recovered from the Arcadia Formation. It is based on an incomplete skull (35 mm long) strongly reminiscent of prolacertids, a group of primitive archosauromorphs that include terrestrial and aquatic relatives of crocodiles and dinosaurs. *Kadimakara* is thought to have been primarily insectivorous and had a maximum body length of ~350 mm.

A possible true lizard or lepidosaur from the Arcadia Formation, *Kudnu mackinlayi*, is represented by a partial skull (<20 mm in maximum length) belonging to a tiny animal ~200 mm long. Possible lepidosaurid or archosauromorph footprints have been reported from the Upper Triassic Blackstone Formation of the Rhondda Colliery near Dinmore in south-eastern Queensland. These tracks (~200 mm long) were attributed to the ichnotaxon *Plectroperna* (known from North America) and probably represent a small quadruped ~1 m in length.

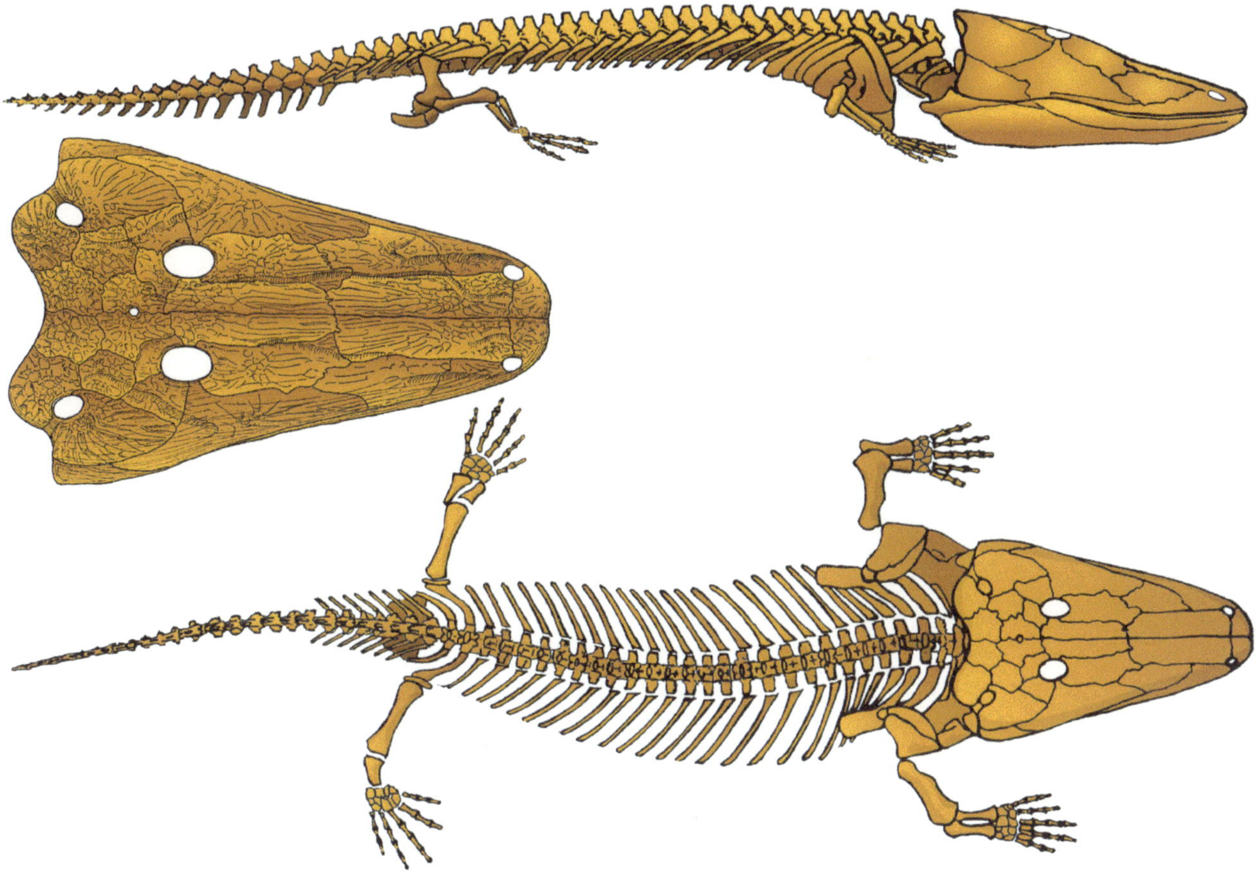

Reconstructed skeleton and skull (after Watson 1958) of the mastodonsaurid *Paracyclotosaurus davidi* from the Middle Triassic Ashfield Shale of St Peters in suburban Sydney.

Early archosaurs

Archosauromorphs – the group that includes dinosaurs, pterosaurs, crocodilians and birds – achieved global distribution during the Triassic and constituted the top predators in many continental faunas. Australian Triassic archosauromorph remains have been described from several deposits including the Arcadia Formation and the Induan-Olenekian Knocklofty Formation (Tasmania Basin) of suburban Hobart. At least two taxa are represented: *Kalisuchus rewanensis,* based on a variety of fragmentary bones including components of the skull, vertebrae, limb and girdle elements recovered from the Arcadia Formation; and *Tasmaniosaurus triassicus*, established on a near-complete skeleton from the Knocklofty Formation. Both *Kalisuchus* and *Tasmaniosaurus* are allied with the Proterosuchidae, a family of primitive archosauromorphs (up to 3 m long) best known from South Africa, India, Russia and China. Proterosuchids are thought to have been semi-aquatic crocodile-like predators that fed on fish, small amphibians and reptiles. Indeed, bone fragments of an unidentified small temnospondyl were found associated with the body cavity of *Tasmaniosaurus* and may represent the last meal of that animal.

The skeleton of *Kalisuchus* displays a number of distinctive proterosuchian features including teeth set in shallow sockets within a continuous groove, a long neck and crocodile-like ankle joint. *Tasmaniosaurus* has a long curved snout with about 16 teeth, and vertebral centra characteristic of large aquatic proterosuchians such as *Proterosuchus* from South Africa and China.

Proterosuchian vertebrae (up to 150 mm high) have been reported from the Olenekian Bulgo

Remains of Triassic reptiles from the Arcadia Formation of Queensland. (Top) Side views of (left) a jaw with teeth and (right) the orbital region of a skull and mandible of the possible lepidosaur *Kudnu mackinlayi*. (Bottom) The parietal region of a skull in (left) dorsal view and (right) side view of the jaws of the prolacertiform *Kadimakara australiensis*.

Sandstone (Narrabeen Group) at Long Reef in the northern suburbs of Sydney. Unfortunately these remains are non-diagnostic, but they demonstrate that new archosauromorph taxa await discovery in the Lower Triassic deposits of Australia.

Dinosaurs

Australian Triassic dinosaurs are virtually unknown. Previous records of a prosauropod, *Agrosaurus macgillivrayi*, purportedly collected in 1844 from the Cape York Peninsula by the crew of HMS *Fly* are incorrect. These specimens have been shown to belong to the prosauropod *Thecodontosaurus* and are actually derived from Triassic deposits of Somerset in England.

The only definitive Australian Triassic dinosaur occurrences comprise a number of carnivorous theropod footprints from Rhaetian (uppermost Triassic) beds in the Callide Basin and the Upper Triassic Blackstone Formation of the Rhondda Colliery, both in south-eastern Queensland. These include a series of large tracks (460 mm long) resembling the widespread Mesozoic ichnotaxon *Eubrontes* and correspond to a carnivorous dinosaur ~6 m in length.

Therapsids

Therapsids (mammal-like reptiles) were one of the most ubiquitous elements of Triassic terrestrial faunas. Their remains are known worldwide and often form the dominant component in most assem-

Footprint attributed to a lizard-like reptile (*Plectroperna*), from the Upper Triassic Blackstone Formation of the Rhondda Colliery near Dinmore, Queensland.

blages. Unusually, therapsid remains are extremely rare in Australia with only a few fragmentary specimens described from the Arcadia Formation of Queensland. Most of this material is attributed to dicynodonts. These bulky herbivores were up to the size of a cow and had stout sprawling limbs, a short tail and a large head with paired tusks and a horny beak, presumably for digging and cropping low-

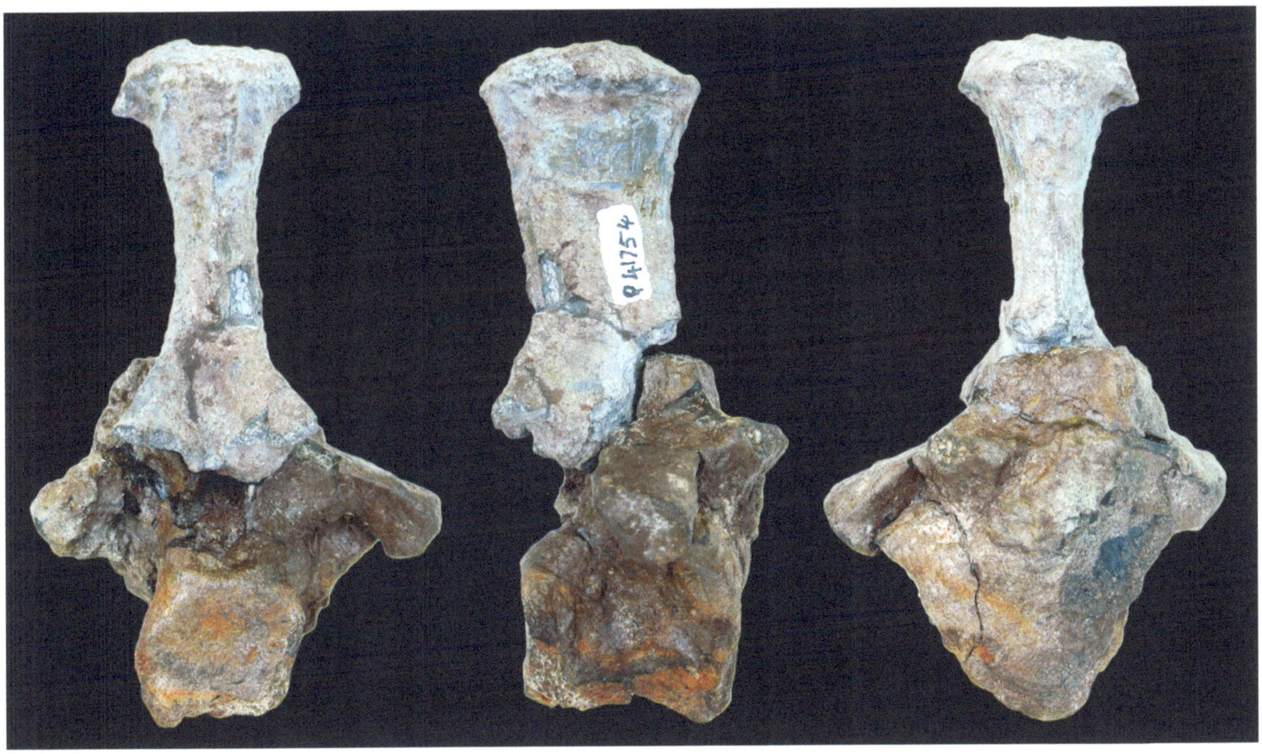

An isolated vertebra (in posterior, lateral and anterior views) attributed to an unknown proterosuchid, from the Lower Triassic Bulgo Sandstone of Long Reef, north of Sydney.

A dicynodont quadrate (ventral articular view), from the Lower Triassic Arcadia Formation of Queensland.

growing vegetation. The Australian dicynodont remains comprise skull fragments comparable in structure to those of the cosmopolitan kannemeyeriid (the group including all Triassic dicynodonts) *Lystrosaurus*, an index fossil for Early Triassic deposits throughout Gondwana (South Africa, Antarctica, India). A series of pentadactyl footprints potentially attributable to a dicynodont have been reported from the earliest Triassic Coal Cliff Sandstone (Narrabeen Group) at the Bellambi Colliery in the southern Sydney Basin. These were named *Dicynodontipus ballambensis* and represent a *Lystrosaurus*-like animal ~850 mm long with a sprawling lizard-like gait.

Two small vertebral centra and the posterior portion of a skull, possibly attributable to a cynodont, have been recovered from the Arcadia Formation. Cynodonts were superficially dog-like carnivores that are closely related to mammals. Little information can be gleaned from the specimens and thus a more definitive assignment cannot be given.

4

BEFORE THE OCEAN: ANIMALS AND PLANTS OF THE JURASSIC

During the Jurassic, Australia was a stable continental area (between 35°S and 65°S palaeolatitude) situated at the south-eastern extremity of the Gondwanan landmass. Its north-western coastal margin bordered the vast Meso-Tethys Ocean, while in the south-west a complex rift valley system separated it from India. The Australian eastern coast comprised the tectonically active Queensland Plateau and Lord Howe Rise. This volcanic region extended to the south, where it incorporated the Western Province of New Zealand and the western coastal terrain of New Caledonia as a shallow near-shore marine margin. The exposed continental landsurface persisted to New Guinea, which then formed the offshore shelf of the Australian region.

Sedimentation in Australia during the Jurassic was intermittent and largely terrestrial in origin, with major river systems occupying lowland regions of the Clarence-Moreton, Surat and Eromanga basins in southern Queensland and northern New South Wales. By the Early Jurassic (Pliensbachian, 185 million years ago), these had expanded into a vast network of coal swamp and lacustrine areas with alternating high-energy/low-energy stream and terrestrial conditions. Upland environments existed in central, western and north-western Australia and provided the principal source of sediments for fluvial deposits throughout most of the Jurassic. Mountainous areas were also present in Tasmania and north-eastern Queensland.

Marine incursions during the Jurassic were generally limited; however, a progressive rise in global sea-levels, which peaked during the Middle Jurassic (Bajocian, ~170 million years ago), formed large shallow marine embayments in the Perth, Carnarvon and Canning basins of Western Australia. Extensive coal swamps also appeared in the rift valley system between Australia and India, suggesting that wetter conditions, possibly tending towards a monsoonal climate with a distinct dry season, had started to develop.

During the Late Jurassic (Kimmeridgian-Tithonian, 155–145 million years ago), marine conditions existed in the north-western Canning Basin and rifting commenced between Australia and Antarctica. This formed a system of deep valleys with lakes, rivers and forests that persisted into the Early Cretaceous.

Australia's best-documented Jurassic fossil-bearing strata are of Lower (Pliensbachian) through to Middle Jurassic (Callovian) age. Sporadic Upper Jurassic marine and continental deposits are known, but these are usually poorly fossiliferous and generally restricted to the Jurassic-Cretaceous boundary.

A characteristic Lower Jurassic sequence is the fluviatile-lacustrine Evergreen Formation of south-eastern Queensland. This unit crops out mainly in the southern Surat Basin and has yielded several important vertebrate finds. The Australian Middle Jurassic fossil record is largely restricted to

Excavation site of the giant temnospondyl amphibian *Siderops*, found in ironstone-rich Lower Jurassic lake deposits of the Evergreen Formation in south-central Queensland. Image courtesy Anne Warren.

sediments of Bajocian-Bathonian age. Probably the most significant are the Bajocian Newmarracarra Limestone (Perth Basin), which has produced a wealth of well-preserved marine invertebrates and represents an important palaeobiogeographical correlation point for the Australian Middle Jurassic, and the Bathonian to Callovian Walloon Coal Measures (Clarence-Moreton Basin/Surat Basin), which constitute one of the most important sources of Australian Jurassic terrestrial plants and rare vertebrate remains (principally footprints).

A rich but relatively geographically restricted Upper Jurassic deposit is the Talbragar Beds. These are exposed in the Talbragar River valley of central-northern New South Wales and have yielded a diverse range of fossil fish, invertebrates and plants that lived in and around a shallow lake ~160 million years ago. The Talbragar Beds have been seriously depleted by years of systematic collecting and are now protected as a national geological monument.

Significant Australian Jurassic deposits and localities

Some of the oldest Australian Jurassic sequences, such as the Evergreen Formation, occur in Queensland and were derived from low-energy river systems, lakes and swamp environments that spread across central and eastern Australia during the later Early Jurassic. The Evergreen Formation comprises a series of basal quartzose sandstones grading upwards into siltstones and shales that are sporadically distributed throughout the Mulgildie, Eromanga and principally Surat basins of south-eastern Queensland. The characteristic lithology represents a progressive decline from high-energy braided stream (characterised by the underlying Precipice Sandstone) to low-energy meandering rivers, floodplains and lakes. Extensive ironstone beds (Westgrove Ironstone Member) in the uppermost part of the unit may indicate marine input through an easterly connection to the open sea (via

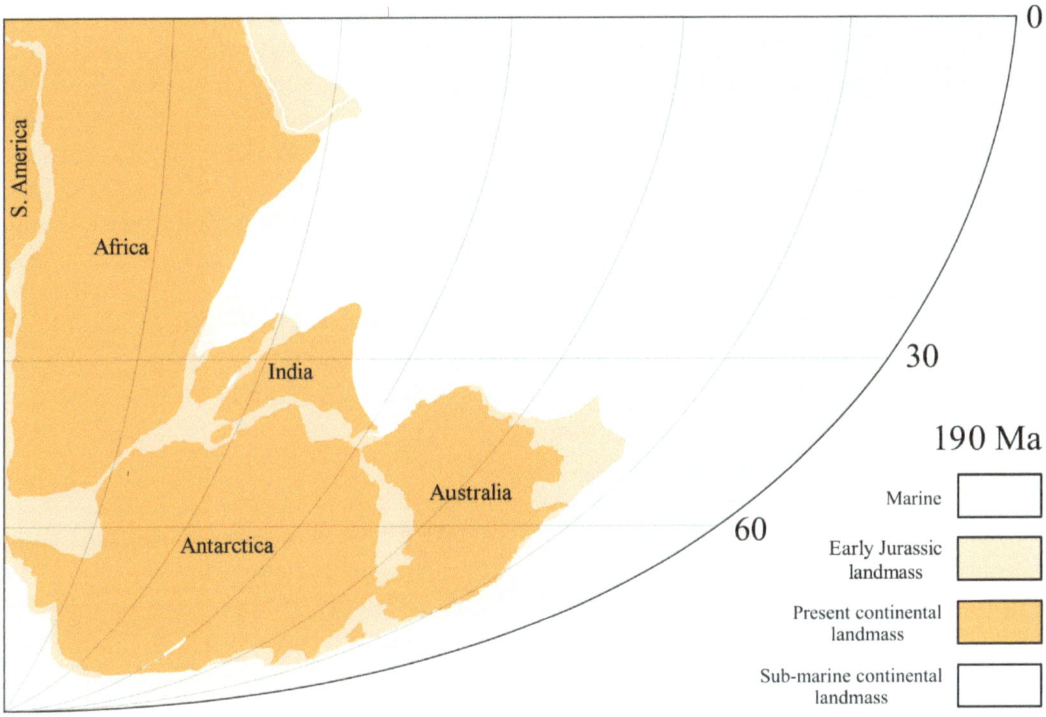

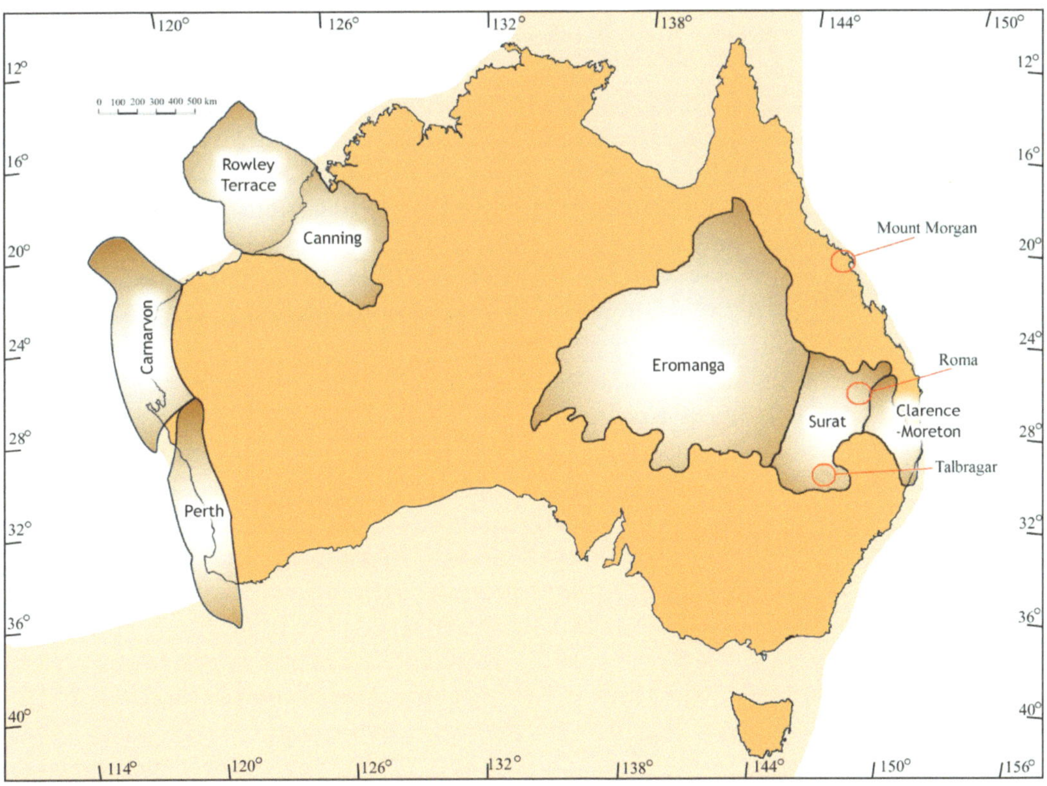

(Top) The geographic position of the Australian continent during the Early Jurassic (Liassic), ~190 million years ago. (Bottom) Principal Jurassic fossil localities (indicated in red) occur mainly throughout the Eromanga, Surat and Clarence-Moreton basins of eastern Australia and in the Perth, Carnarvon and Canning basins and offshore Rowley Terrace of Western Australia.

the Clarence-Moreton Basin). The age of the Evergreen Formation has been determined as Pliensbachian-Toarcian using plant microfossils.

Probably the best-known of Australia's Middle Jurassic marine rocks belong to the Newmarracarra Limestone, a richly fossiliferous sequence of shallow near-shore marine strata that crops out near the township of Geraldton (Perth Basin) on the south-western coast of Western Australia. The Newmarracarra Limestone conformably overlies a series of slightly older marine units including the Colalura Sandstone, which is notable because it has produced isolated vertebrate remains (marine reptiles and dinosaurs). Bivalves and cephalopods dominate the Newmarracarra Limestone fossil assemblage. They occur within a greyish-yellow limestone composed mainly of bivalve shell hash. The Newmarracarra Limestone mollusc fauna, particularly ammonites, has been used to demonstrate a middle Bajocian age for the unit.

Australian Middle Jurassic terrestrial fossils are recorded from several units, most notably the Walloon Coal Measures of south-eastern Queensland. These deposits comprise a series of volcanolithic sandstones, coal, mudstones and siltstones that extend over a broad area of the Surat and Clarence-Moreton basins. The Walloon Coal Measures have a maximum thickness of ~250 m, and have produced fossils from surface road cuttings and underground coal-mining operations. The Walloon Coal Measures are considered Bathonian to Callovian in age on the basis of plant macrofossil and spore/pollen fossil assemblages. Their depositional setting was a combination of major channel, flood basin and coal-forming peat swamps generated by river systems draining to the south.

Although tradionally regarded as Lower Jurassic, the Talbragar Beds are now thought to be Upper Jurassic in age. They correspond to a lacustrine setting, probably part of an extensive freshwater lake system that was smothered by a volcanic eruption. The main fossil-bearing horizon comprises an ochre-coloured iron-rich cherty shale lens that is weathered to loose blocks floating in soil. The individual rock slabs may be displaced erosional remnants of the original lakebed deposit (regarded as part of the underlying Purlawaugh Formation); however, recent work has uncovered fresh exposures. The principal outcrop occurs on the north-eastern slope of Farrs Hill, north-east of Gulgong in northern New South Wales. None of the slabs from these beds have a thickness greater than 600 mm, suggesting that the original layer was very thin and may have accumulated over a relatively short period (in accordance with the volcanic scenario). No definitive microfossil dating has been established for the Talbragar Beds; their Late Jurassic age has been proposed on the basis of the macrofossil assemblage and isotopic dating of zircon crystals.

Land plants

Australian Jurassic plant macrofossils have been recorded from deposits of Lower to Upper Jurassic age. These floras generally resembled communities from elsewhere in the Gondwanan region, with araucarian and podocarp conifers dominating the canopy and cycadophytes (cycads and relatives), ginkgoaleans (ginkgos), pteridophytes (ferns), equisetaleans (horsetails), bryophytes (mosses and liverworts) and lycophytes (clubmosses) as understorey. Flowering plants, which did not diversify until the Cretaceous, were either absent or insignificant components of the assemblage. Lower Jurassic macrofloras from Western Australia and Queensland suggest that some palaeolatitudinal variability was evident between the dominant vegetative types. For example, lower-latitude (25–35°S palaeolatitude) floras from Western Australia, which are characterised by material from the Cockleshell Gully Formation of the Perth Basin, reflect the influence of warm equatorial currents from the Meso-Tethys Ocean. In contrast, eastern assemblages such as the Precipice Sandstone and Evergreen Formation of the Surat Basin in Queensland, correspond to higher-latitude (40–75°S palaeolatitude) more-seasonal vegetation.

By the Middle Jurassic, uniform floras had developed across the Australian continent and throughout much of Gondwana. Patterns of latitudinal zonation evident during the Early Jurassic had become less pronounced, possibly in response to rising global sea levels and the onset of cooler but more equable climatic conditions. Australian Middle Jurassic macrofloras are well-known from the Bajocian-Bathonian high-latitude (~60°S

Foliage of various Jurassic plants from Queensland. (Clockwise from top left) The horsetail *Equisetum* (Walloon Coal Measures), frond of the bennettitalean *Otozamites* (Walloon Coal Measures), foliage of the podocarp *Elatocladus* (Stanwell Coal Measures) and the fern *Ptylophyllum* (Injune Creek Beds).

palaeolatitude) humid swamp communities of the Walloon Coal Measures. These preserve evidence of a rich understorey comprising osmundacean, dicksoniacean and dipteridacean ferns, liverworts, lycophytes, equisetaleans, the pentoxylacean *Taeniopteris* (which persisted into the Cretaceous) and bennettitaleans (primitive cycad-like plants). Araucarian (*Podozamites*) and podocarp (*Elatocladus*) conifers formed the principal canopy elements. Well-defined growth rings in preserved conifer wood is suggestive of seasonality. Extensive accumulations of peat indicate that any drought conditions must have been short-lived.

The Upper Jurassic Talbragar Beds of New South Wales have yielded evidence of a diverse vegetation comprising equisetaleans, ferns, bennettitaleans such as *Otozamites*, pentoxylaceans (*Taeniopteris*) and a variety of podocarp (*Rissikia talbragarensis, Elatocladus australis*) and araucarian (*Podozamites jurassica, Brachyphyllum*) conifers. These elements have been described as a Jurassic kauri pine forest with an araucarian-dominated canopy interspersed with a pentoxylacean-rich lower tree zone, scattered podocarps and fern-cycad heath zones surrounding lake margins. Correlations based on spores and pollen, together with macrofossils recovered from Jurassic-Cretaceous

(Right) Foliage of the araucarian conifer *Podozamites jurassica* from the Talbragar Beds of central New South Wales. (Left) An example of the living Queensland kauri pine, *Agathis robusta*.

boundary strata in the Gippsland Basin of southern Victoria and Perth/Canning basins of Western Australia, show that these basic Jurassic floral elements (araucarian/podocarp conifers, pentoxylaceans) were persistent across the continent throughout the remainder of the period.

Marine and non-marine invertebrates

The current record of Australian Jurassic macroinvertebrates is derived almost entirely from the marine deposits of Western Australia. Nevertheless, some rare insects and freshwater bivalves including *Prohyria scepticsi* and *Prosphaerium talbragarensis* are known from various localities, such as the Pliensbachian-Toarcian Evergreen Formation of Queensland and the Cockleshell Gully Formation of Western Australia, undifferentiated Lower Jurassic strata near Korumburra in Victoria and the Upper Jurassic Talbragar Beds in New South Wales.

Australian Early Jurassic marine macroinvertebrates are essentially unknown. However, highly diverse Middle Jurassic assemblages have been

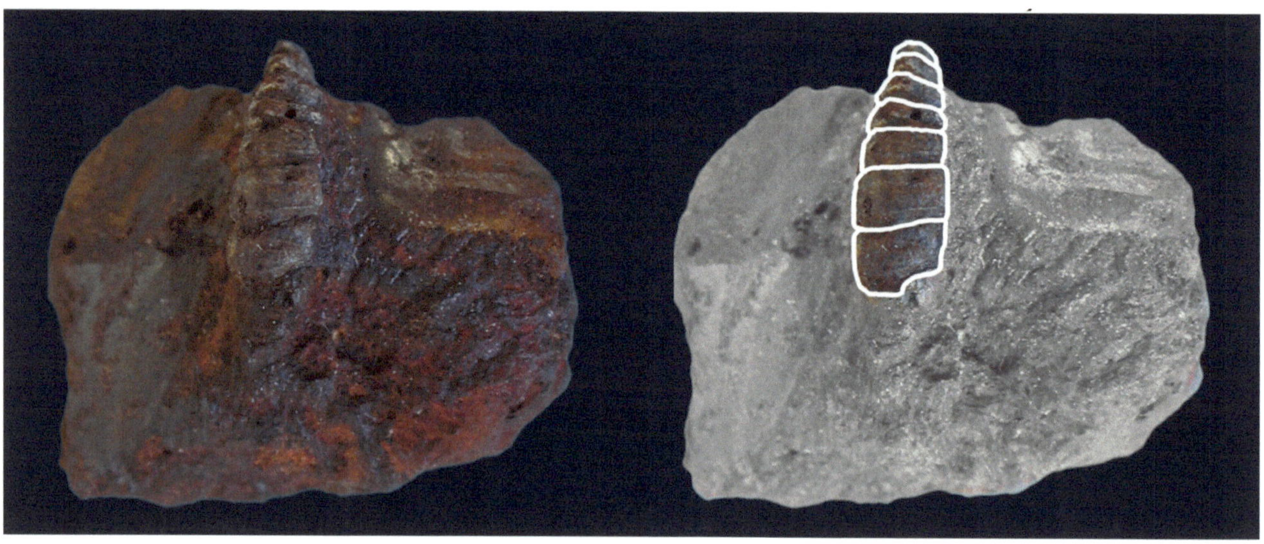

Segments of a delicate fossil millipede *Decorotergum warrenae*, preserved in ironstone-rich sediments of the Evergreen Formation of south-eastern Queensland.

identified from several units including the lower Bajocian Newmarracarra Limestone of the Geraldton area in south-western Western Australia. More than 60 mollusc genera and species (mainly bivalves and ammonites) have been documented, with most representing Tethyan or cosmopolitan lineages. Almost all the species-level taxa are endemic, perhaps suggesting that the Newmarracarra Limestone fauna was isolated within a narrow embayment extending south-west between the Indian and Australian portions of the Gondwanan landmass.

The Newmarracarra Limestone bivalves are dominated by shallow water forms including infaunal or burrowing trigonids (*Trigonia*), hard surface cementers (e.g. the oysters *Astarte* and *Gresslya*) and free-swimming pectinids (scallops such as *Pecten*). Pelagic cephalopods comprise nautiloids (*Nautilus*), squid-like belemnites (*Dicoelites*) and ammonites. The latter incorporate a mix of cosmopolitan (e.g. *Fontannesia, Otoites, Sonnonia*), circum-Pacific (e.g. *Pseudotoites*) and endemic genera (*Newmarracarroceras*), indicating palaeobiogeographical links to contemporary assemblages in South-East Asia

Marine invertebrates from the middle Jurassic Newmarracarra Limestone. (Clockwise from top left) The bivalves *Lopha marshii* and *Cucullaea semistriata,* the ammonite *Newmarracarroceras clarkei* and the distinctive ridged bivalve *Trigonia moorei.*

Jurassic fish fossils from the famous Talbragar Beds in central-eastern New South Wales. (Top) The early teleost *Cavenderichthys talbragarensis*. (Bottom) The endemic archaeomenid *Archaeomene tenuis*.

(Indonesia) and the west coast of the Americas (North America, Argentina).

Australian Late Jurassic marine macroinvertebrates are best known from the Oxfordian-Tithoinian Alexander Formation and Jarlemai Siltstone of the Edgar Ranges, near Broome in north-western Western Australia. The fauna includes indeterminate ophiuroids (brittle-stars), together with distinctive belemnite (*Belemnopsis*), ammonite (*Kossmatia*, *Virgatosphinctes*) and bivalve (*Buchia*, *Calpionella*) taxa. These are comparable to species already described from South-East Asia (Indonesia) and they support interpretation of Western Australian Late Jurassic assemblage as part of an essentially Tethyan faunal extension into the eastern Gondwanan region.

Non-marine fish

Australian Jurassic fish fossils are presently only reported from non-marine deposits. Fragmentary material has been recovered from the Bathonian Walloon Group of the Ipswich Coal Fields near Brisbane, and the Hutton Sandstone at Monto in south-eastern Queensland. These specimens are largely indeterminate actinopterygians (ray-finned fish), with the exception of a body section from Monto.

Uarbryichthys latus from the Jurassic Talbragar Beds of central-eastern New South Wales.

This shows impressions of thick rhombic scales similar to those of semionotiforms, primitive ray-finned fishes widespread during the Jurassic.

A beautifully preserved Australian Late Jurassic fish fauna is known from the Talbragar Beds of central-eastern New South Wales. This deposit has produced a vast number of complete skeletons representing a variety of endemic Gondwanan taxa such as the archaeomenids (a family restricted to Australia), *Archaeomene tenuis* and *Madariscus robustus* and a macrosemiiform, *Uarbryichthys latus*. Also present is a species of the globally distributed palaeoniscoid *Coccolepis*, *C. australis*, and the oldest Australian teleost (modern ray-finned fish), *Cavenderichthys talbragarensis*. Unidentified coelacanth (Crossopterygii), amiiform (bowfins, primitive actinopterygians still living today) and hybodontiform shark remains, together with the enigmatic archaeomenid *Aphnelepis australis*, represent additional species with uncertain relationships.

Amphibians

Jurassic temnospondyl amphibian remains have been recorded from several deposits in Australia. *Austropelor wadleyi* was named in 1941 on the basis of a mandible fragment from the Lower Jurassic Marburg Formation on the banks of the Brisbane River in south-eastern Queensland. The specimen shows a distinctive arrangement of tooth sockets (decreasing in size towards the front of the jaw) characteristic of brachyopoids, a temnospondyl group known from the Triassic to Cretaceous of central Asia, India, South America, South Africa and Australia.

A second fossil amphibian, *Siderops kehli*, has been described on the basis of a near-complete skeleton (missing only parts of the limbs and tail) from the Pliensbachian-Toarcian Evergreen Formation (Westgrove Ironstone Member) of south-eastern Queensland. This huge (2.5 m) crocodile-like aquatic temnospondyl represents a derived member of the Chigutisauridae (a brachyopoid family). It is distinguished by a broad head (nearly 650 mm across) with upward-facing orbits and large lance-shaped teeth along the side of the jaw. *Siderops* is thought to have lived in freshwater rivers and lakes alongside non-marine plesiosaurs.

Plesiosaurs

Plesiosaurs (Plesiosauria) were a group of Mesozoic marine reptiles characterised by a long neck, box-like body and four propulsive flippers. Two distinct

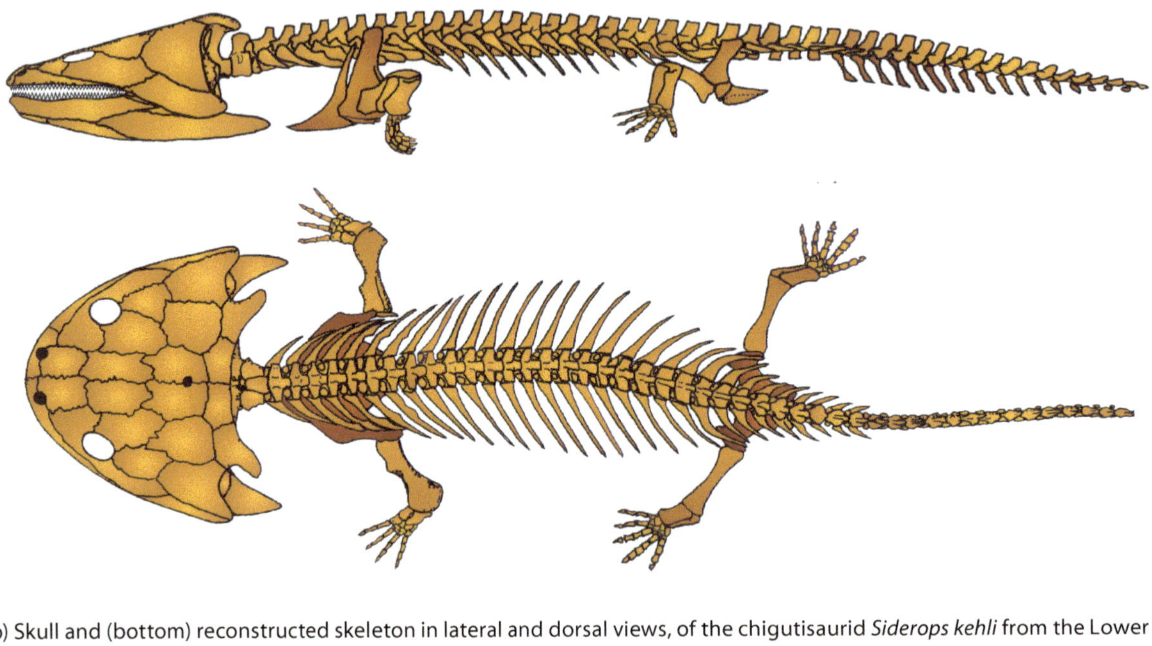

(Top) Skull and (bottom) reconstructed skeleton in lateral and dorsal views, of the chigutisaurid *Siderops kehli* from the Lower Jurassic Evergreen Formation of south-eastern Queensland. Reconstruction after Warren and Huchinson (1983).

Reconstruction of the giant Early Jurassic temnospondyl *Siderops kehli* based on an articulated skull and skeleton from the Evergreen Formation of Queensland.

functional body-types have been recognised: the small-headed long-necked plesiosauromorphs, and large-headed shorter-necked pliosauromorphs. Both were globally distributed throughout the Jurassic and Cretaceous; however, Jurassic plesiosaur fossils are extremely rare in Australia. The oldest recorded remains include a fragmentary skeleton of a small pliosauromorph from the lowermost Jurassic Razorback Beds, near Mt Morgan in Queensland. This specimen shows some similarities in vertebral morphology to the primitive non-marine rhomaleosaurid pliosauromorphs such as *Bishanpliosaurus youngi* from the Lower Jurassic (Liassic) of China. The Mt Morgan specimen has not yet been examined in detail, and its definitive relationships are unclear.

Two other extremely fragmentary plesiosaur fossils have been reported from the Evergreen Formation near Wandoan in south-eastern Queensland. This material consists of associated limb girdle and vertebral elements that may be referable to the plesiosauromorph family Plesiosauridae.

Several isolated plesiosaur vertebrae and a finger bone (phalanx) have been excavated from the Middle Jurassic (middle Bajocian) Colalura Sandstone of Bringo railway cutting near Geraldton in Western Australia. One of these specimens (a dorsal vertebra) was considered a probable elasmosaurid and thus represents the oldest example of this group from Australia.

Dinosaurs

Australian Jurassic dinosaurs are known from sporadic body fossils and footprint evidence. Probably the most famous discovery is the incomplete skeleton (comprising most of the hindleg, partial pelvis and vertebral series) of *Rhoetosaurus brownei*, a primitive sauropod (*Brontosaurus*-like herbivores) from the Bajocian (Middle Jurassic) Hutton Sandstone near Roma in central-southern Queensland. *Rhoetosaurus* shares a range of diagnostic characters, in particular the presence of rudimentary pleurocoels (cavities in the vertebrae thought to help reduce weight without compromising structural strength), with cetiosaurids, a group of archaic sauropods well-known from the Lower-Middle Jurassic of England, North Africa, North and South America and especially China. *Rhoetosaurus* has been specifically allied with the Shunosaurinae, cetiosaurids that possessed short stiffened tails with a bony club at the end. The tail vertebrae of *Rhoetosaurus* are very similar to those of shunosaurines from China, suggesting that a defensive tail club might have been present in this Australian sauropod.

Isolated Middle Jurassic (Bajocian) dinosaur bones have been found in the marine Colalura Sandstone of the Perth Basin in Western Australia. These represent the remains of carcasses that were washed out to sea from nearby coastal environments. Specimens include an indeterminate sauropod tail vertebra from the Bringo railway cutting, near Geraldton. This specimen is not diagnostic but it displays similar proportions to the tail vertebra of cetiosaurids. A partial theropod (carnivorous dinosaur) tibia (shinbone), also from the Bringo railway cutting, has been named *Ozraptor subotai*. This intriguing taxon differs from all other

Many plesiosaur fossils from Jurassic non-marine deposits have been attributed to rhomaleosaurids. These fast-swimming predators probably hunted fish and other marine reptiles. Fragmentary plesiosaur remains from the Lower Jurassic of Queensland are probably referable to this group and represent animals that inhabited a brackish estuary system ~180 million years ago.

JOSH.

An enigmatic series of vertebrae attributed to an unknown plesiosaur from the Lower Jurassic Evergreen Formation at Wandoan in south-eastern Queensland. The articulated cervical series is exposed in lateral view.

The reassembled foot of the Jurassic sauropod dinosaur *Rhoetosaurus brownei* as previously displayed in the Queensland Museum. Image courtesy Thomas Rich.

Reconstruction of the primitive Middle Jurassic sauropod dinosaur *Rhoetosaurus brownei* from the Hutton Sandstone near Roma in Queensland. The incomplete postcranial skeleton of *Rhoetosaurus* is still the only Australian Jurassic dinosaur body fossil known from anything more than an isolated bone.

known theropod groups and cannot yet be assigned to any currently recognised family.

Australian Lower Jurassic dinosaur footprints are known from the Precipice Sandstone and Razorback Beds of southern Queensland. These have yielded tracks of large and small (*Anomoepus*) ornithopods (bipedal plant-eaters) and indeterminate theropods.

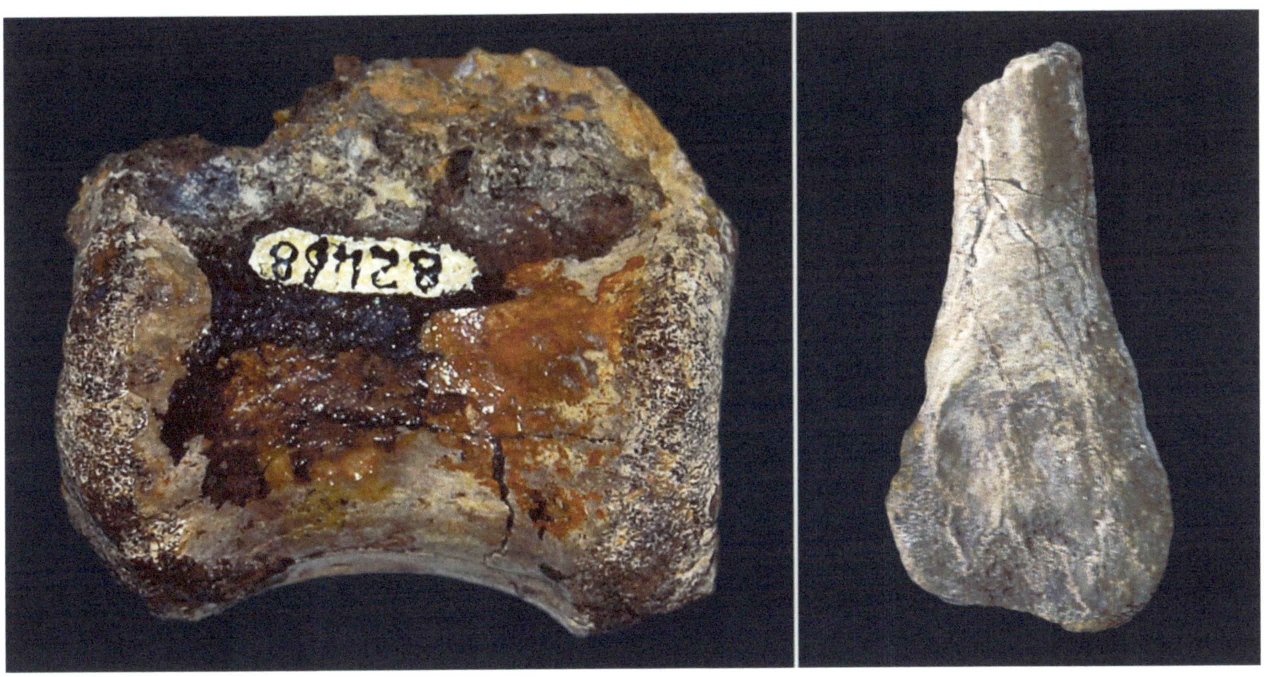

Middle Jurassic dinosaur fossils from the Colalura Sandstone near Geraldton in Western Australia. (Left) An isolated caudal vertebra attributed to a sauropod. (Right) The distal tibia of *Ozraptor subotai*. Image courtesy John Long.

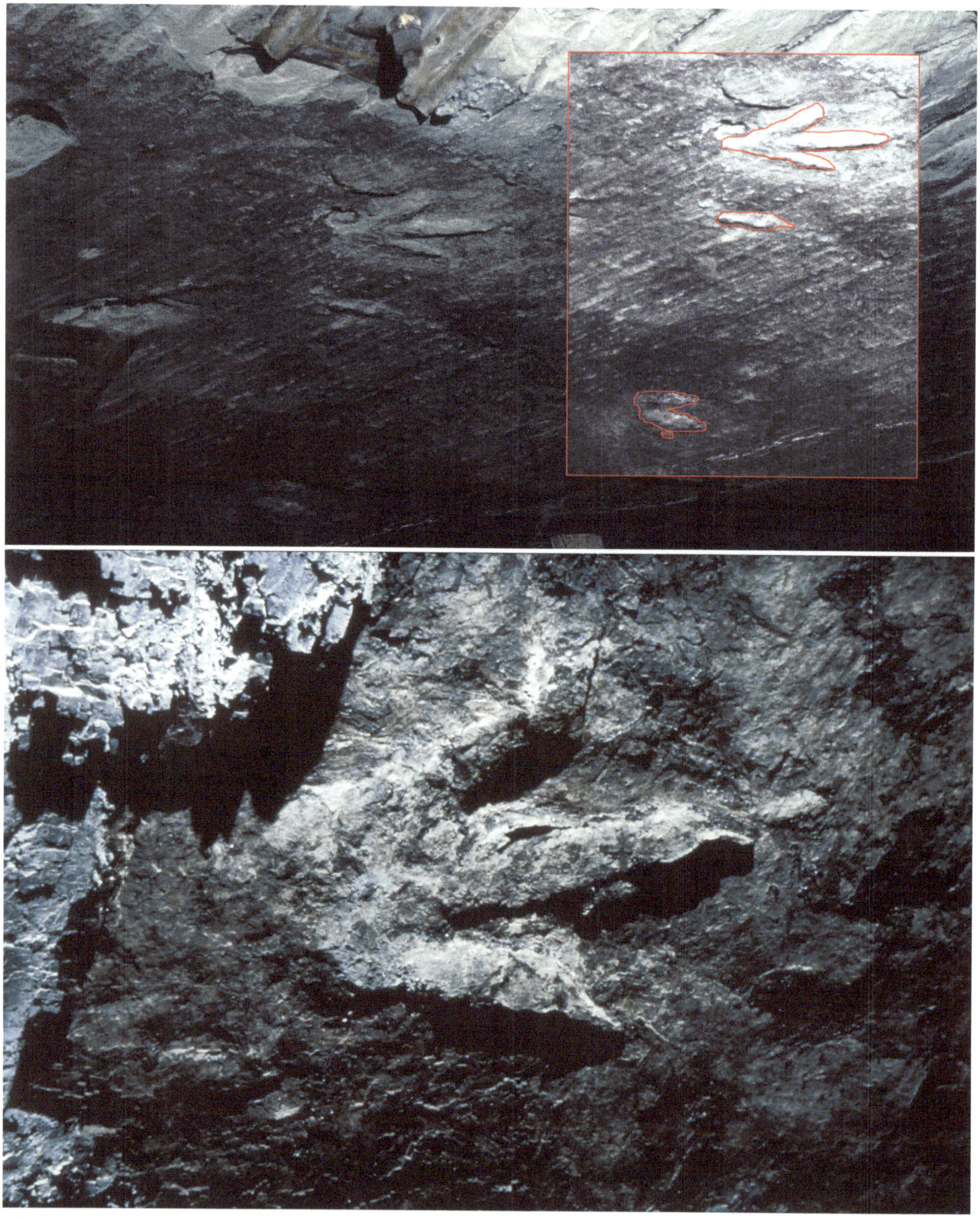

Dinosaur tracks preserved in the roof of a coal mine in the Middle Jurassic Walloon Group Coal Measures near Balgowan in south-eastern Queensland. The tracks were exposed when the underlying coal and siltstone layers were removed during tunnelling. Images produced by Tim Hamley and courtesy Thomas Rich.

Extensive Middle Jurassic (Bajocian-Bathonian) dinosaur trackways have been found in the Walloon Group Coal Measures near Balgowan in south-eastern Queensland. These indicate a range of ornithopods (*Changeipus bartholomaii*), a quadrupedal thyreophoran and theropods. The latter include small and large (*Eubrontes*) forms up to 710 mm in length, suggesting large carnosaurs with body lengths of up to 12 m.

5

THE GREAT INLAND SEA: AUSTRALIA IN THE EARLY CRETACEOUS

The Cretaceous began around 145 million years ago (Berriasian), when the Australian landmass comprised an easternmost high-latitude (50°–85°S) extremity of the Gondwanan supercontinent. Its western margin formed a continuous landsurface with India and its eastern edge comprised the upland Queensland Plateau and Lord Howe Rise. As in the preceding Jurassic, most of the continental area remained landlocked.

By the Valanginian (140 million years ago), narrow marine straits had developed in the expanding rift systems established next to India in the west and Antarctica in the south. A rise in global sea-levels during the Barremian (130 million years ago), coupled with downwarping of the Carpentaria, Eromanga and Surat basins, prompted a major marine invasion into central-eastern Australia. This was accompanied by marine transgressions in the north-western Carnarvon Basin and the coastal margin of the Perth Basin, which opened new areas of shallow epicontinental marine habitat. Continued sea-level rises during the Aptian (reaching a peak around 120 million years ago) inundated most of the Australian continent. Island areas persisted in the north-western Kimberley region and the volcanically active eastern Queensland Plateau. The south-eastern continental margin remained continuous with Antarctica. The Aptian inland sea established open ocean connections through the western Perth and Carnarvon basins, and in the south via a broad strait across the Nullarbor Plain. In the east, a vast epicontinental sea extended from the Gulf of Carpentaria into the Great Artesian Basin and across the Northern Territory, with a narrow south-eastern channel in the Surat Basin (around the modern New South Wales/Queensland border).

The onset of cooler climatic conditions in the Aptian pushed sea-surface temperatures close to freezing. Evidence for this has been found in isotopic and sedimentary analyses, which indicate average sea-level palaeotemperatures of ~12°C. Extensive coastal ice build-up appears to have occurred along the southern margins of the inland seaway during prolonged high-latitude winters.

By the early-middle Albian (~115 million years ago), these extreme conditions gave way to warmer, more temperate climates. The inland sea regressed, leaving isolated inundations along the Western Australian and Northern Territory coastlines. Despite this, an extensive epicontinental seaway remained in central-eastern Australia throughout the middle and late Albian (up to ~105 million years ago). By the earliest Late Cretaceous (Cenomanian, ~99 million years ago) this finally retreated northwards towards its open ocean outlet in the Gulf of Carpentaria.

The oldest of Australia's Lower Cretaceous marine fossil-bearing sequences include the Nanutarra Formation and Birdrong Sandstone. These units are exposed as weathered cliffs in the Murchison River region of north-western Western Australia (Carnarvon Basin), and have produced invertebrate and vertebrate fossils. Much richer

The Moon Plain north of Coober Pedy in South Australia is one of Australia's best-documented Lower Cretaceous marine fossil localities. Its deposits record evidence of near-freezing climatic conditions during the Aptian, ~120 million years ago.

Lower Cretaceous assemblages are known from the vast Aptian inland seaway deposits of the Eromanga and Carpentaria basins, incorporating the Bulldog Shale and Wallumbilla Formation of South Australia and northern New South Wales/southern Queensland respectively. These have yielded some of Australia's most spectacular fossils, especially from the opal-bearing strata around Coober Pedy and Andamooka in South Australia and White Cliffs in New South Wales. Many of the fossils have been totally replaced by opal. They typically occur alongside low-temperature palaeoclimatic indicators (e.g. ice-rafted boulders) in underground and open-cut mine workings.

Marine fossil assemblages from the Albian differ markedly from those found in older Aptian sediments. For example, invertebrates appear to have undergone a significant drop in diversity and marine vertebrates exhibit a rearrangement in faunal composition (e.g. plesiosaurs were replaced by ichthyosaurs and marine turtles). Although the causes are unclear, widespread marine regression coupled with climatic warming might have provided an effective catalyst. The best-known Albian faunas are derived from epicontinental marine deposits of central-eastern Australia. Key units include the Toolebuc Formation and Allaru Mudstone, both of which have produced a wealth of superbly preserved specimens mainly from weathered blacksoil plains around the townships of Boulia, Hughenden and Richmond in central-northern Queensland. This region was dominated by shallow marine environments ~110 million years ago (late Albian), and represents one of the richest Mesozoic fossil sequences yet discovered on the Australian continent.

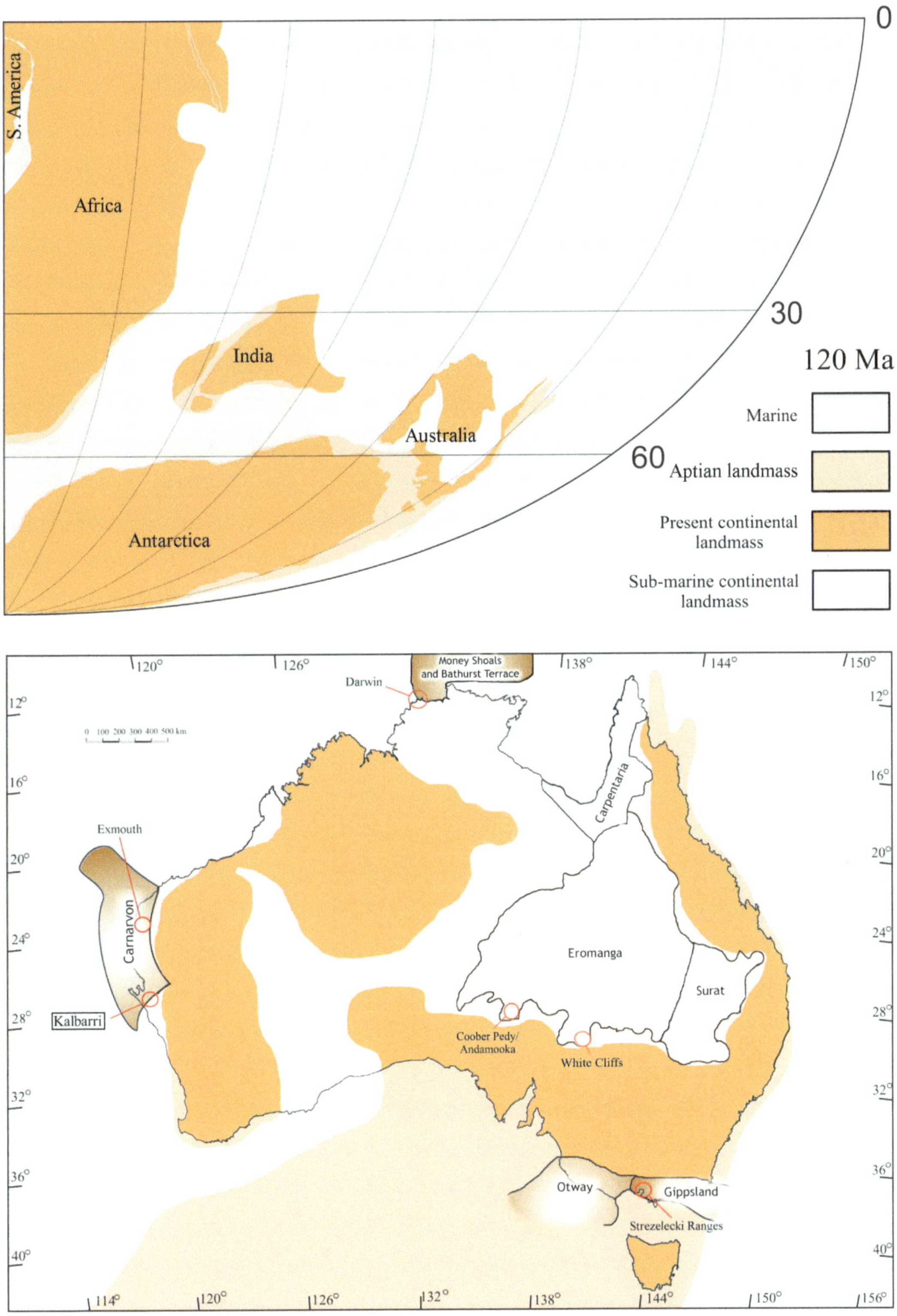

(Top) The geographic position of the Australian continent during the Early Cretaceous (Aptian), ~120 million years ago. (Bottom) Principal Aptian and Hauterivian-Barremian (boxed) fossil localities (indicated in red) are mainly in the Eromanga Basin of eastern Australia, the Carnarvon Basin of Western Australia and the offshore Money Shoals Platform in the Northern Territory.

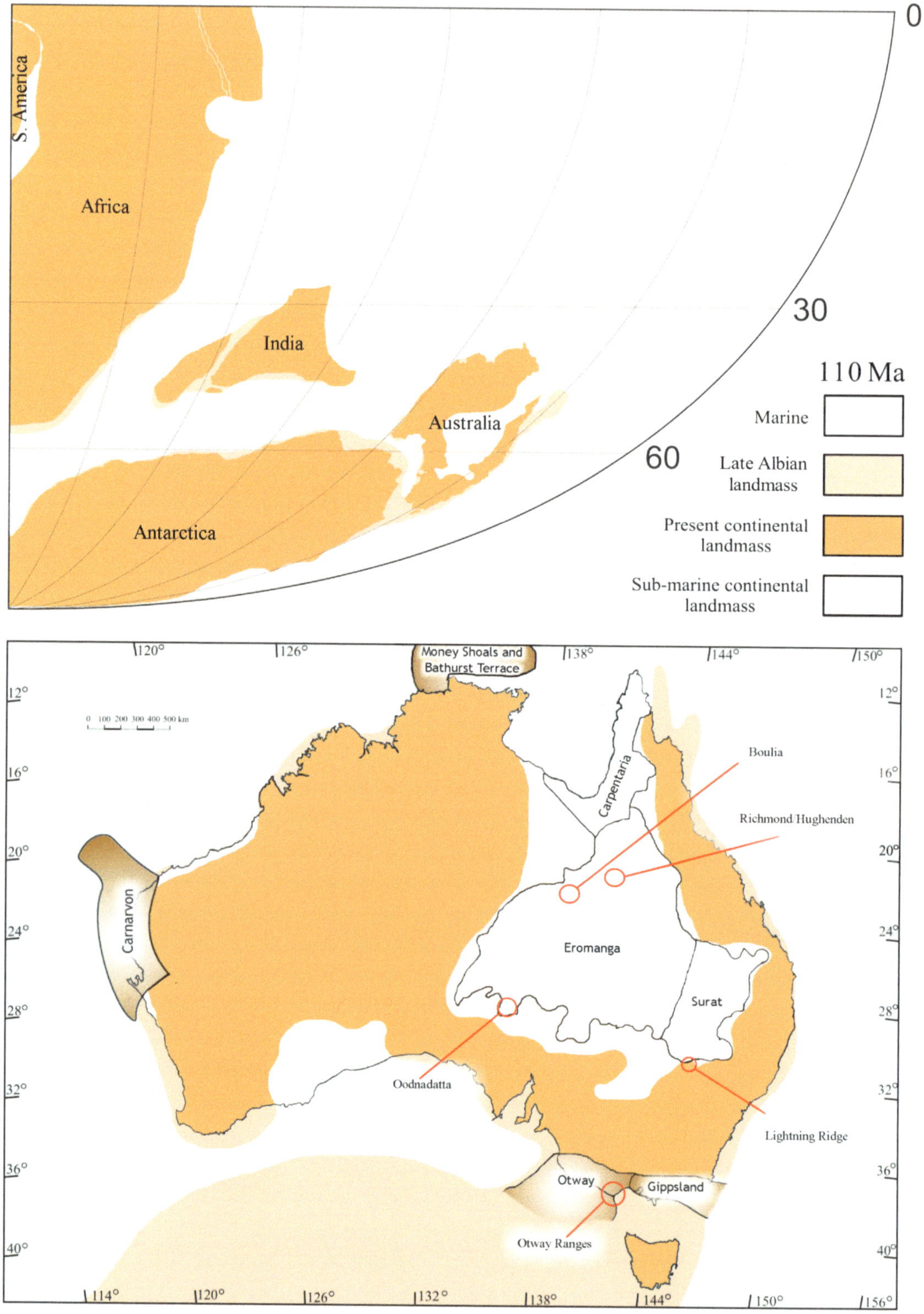

(Top) The geographic position of the Australian continent during the Early Cretaceous (Albian), ~110 million years ago. (Bottom) Albian fossil localities (indicated in red) are best-known from the Eromanga/Surat basins and the Otway Ranges of eastern Australia.

Significant Australian Lower Cretaceous marine deposits and localities

The Nanutarra Formation and Birdrong Sandstone record the initial onset of transgressive shallow marine conditions in the north-western part of the Australian continent during the middle-late Neocomian (Hauterivian-Barremian). These rock units are lateral equivalents that crop out along the north-eastern and southern margins of the Carnarvon Basin respectively. They comprise mainly silty sandstones grading upwards to fine claystone layers in the uppermost part of the sequence. Glauconitic beds are common and both units show evidence of extensive bioturbation (disturbance of sediments by burrowing organisms). The palaeoenvironmental setting is thought to have been shallow coastal marine, with deposition over irregular seafloor topography composed of Palaeozoic basement rocks. Age estimates derived from macrofossils, marine microplankton, fossil spores and pollen indicate that deposition began in the north-western Carnarvon Basin during the early Hauterivian, and proceeded south-eastwards through the Barremian.

Australian Aptian marine rocks can be typified by the Bulldog Shale of South Australia. This unit comprises a series of finely laminated mudstones and shales deposited under transgressive shallow marine conditions. The presence of dark organic-rich layers and low microplankton diversity indicate poorly oxygenated bottom-waters. Characteristically, upper sections of the Bulldog Shale are severely leached and bleached white by weathering. This is best seen in the areas around Andamooka and Coober Pedy in northern South Australia, where bleaching may extend to a depth of approximately 40–50 m and incorporate extensive deposits of opal (hydrated silica, $SiO_2.nH_2O$). The opal at Andamooka and Coober Pedy commonly occurs

The opal mines of Coober Pedy in South Australia have produced some of the most spectacular Lower Cretaceous marine vertebrate and invertebrate fossils found in Australia. The bleached white claystone shown here forms a weathered upper layer of the Aptian Bulldog Shale.

Evidence for ancient climates can be based on a variety of sedimentary structures. (Clockwise from top) Dense arrangements of growth rings (with magnified section) in fossilised tree trunks are indicative of seasonal conditions. Erratic boulders composed of volcanic porphyry or silcrete can be rafted hundreds of kilometres and deposited in fine sea-floor mudstone by floating ice. Devonian gastropods preserved in erratic boulders from Lower Cretaceous marine deposits in central Australia suggest ice-transport occurred via westerly-flowing currents from what is now New South Wales. Glendonites are crystal pseudomorphs of the mineral ikaite, which only forms in freezing sea-floor mud. All specimens shown here are from the Bulldog Shale of South Australia.

An excavation in the Albian Toolebuc Formation deposits near Boulia in western Queensland. Fossils from these deposits typically occur in yellow limestone concretions (foreground) that can be treated with acid to reveal perfectly preserved vertebrate fossils.

along cracks or faults or within cavities left by fossil remains. On occasion, it directly replaces fossilised organic remnants, as evidenced by the occasional preservation of soft tissue structures. Where bleaching is absent, the Bulldog Shale is dominated by shale horizons containing widespread gypsum and fossiliferous carbonate limestone concretions. Macroinvertebrate and fossil plankton assemblages infer an early Aptian to early Albian age. Associated palaeoclimatic indicators including ice-rafted boulders, glendonites (calcite pseudomorphs of the mineral ikaite, which forms only between –1° and 6°C) and wood with dense growth bands also suggest that highly seasonal cool to very cold (possibly with winter freezing) climatic conditions were prevalent.

The Rolling Downs Group comprises the basal Aptian Wallumbilla Formation (laterally equivalent to the Bulldog Shale), overlain by the Albian marine Toolebuc Formation, Allaru Mudstone, Mackunda Formation and the predominantly Upper Cretaceous Winton Formation, which is terrestrial in origin. The Albian deposits comprise mudstones, siltstones, calcareous shale and limestones that are widely distributed throughout the Eromanga Basin in central and northern Queensland. The Rolling Downs Group records the final transgressive phase of the Early Cretaceous inland sea, with low-energy offshore environments (Toolebuc Formation/Allaru Mudstone) grading upwards to high-energy shallow-water conditions (Mackunda Formation). The Rolling Downs Group has been dated using rich microfossil (plankton, spores, pollen) and mollusc assemblages (bivalves, gastropods, scaphopods, cephalopods). These indicate a middle to upper Albian age for the Toolebuc Formation/Allaru

Mudstone, and uppermost Albian to Cenomanian age for the overlying Mackunda Formation and Winton Formation.

Marine invertebrates

The Australian Early Cretaceous marine macroinvertebrate record is very well-documented. Berriasian-Barremian assemblages are known from the Nununtarra Formation, which crops out in the Ashburton River region (Carnarvon Basin) of northwestern Western Australia. The fauna includes bivalves, gastropods, squid-like belemnites and corals (*Actinastraea*). The bivalves *Camptonectes* and *Trigonia*, together with the belemnopsid belemnite *Hibolithes*, are some of the most frequently occurring taxa. They represent Jurassic lineages apparently well-established in the Australian region since the early part of the Mesozoic.

Following the massive expansion of epicontinental marine environments during the Aptian, macroinvertebrate assemblages became extremely diverse. Upwards of 63 benthic mollusc taxa (including gastropods, bivalves and scaphopods), together with cephalopods, echinoderms (brittle-stars, starfish, crinoids and sea urchins), decapod crustaceans, brachiopods (lamp shells) and sponges have been described from the Bulldog Shale in South Australia and the laterally equivalent Wallumbilla Formation of Queensland/northern New South Wales. Most of the identified ammonites (e.g. *Australiceras*, *Sanmartinoceras*, *Tropaeum*), bivalves (e.g. *Eyrena*, *Laevitrigonia*, *Maccoyella*, *Panopea*, *Teredo*), gastropods (e.g. *Euspira*) and scaphopods (e.g. *Dentalium*) represent widespread or globally distributed forms, and infer a northern warm-water origin for their respective lineages. However, the fauna is

Aptian (Lower Cretaceous) marine invertebrates. (Clockwise from top left) An ophiuroid or brittle-star, a starfish, the opalised sponge *Pursiphonia clarkei* and an opalised crinoid calyx attributed to *Isocrinus australis*. All specimens are from the Bulldog Shale of Coober Pedy in northern South Australia.

Aptian (Lower Cretaceous) marine invertebrates. (Clockwise from top left) The ammonites *Australiceras transiente* (Walsh River, Queensland) and *Tropaeum imperator* (Bulldog Shale, South Australia), the bivalves *Eyrena linguloides* and *Cyrenopsis australiaensis* with a magnification of the encrusting bryozoan *Lepralia*, the brachiopod *Rhynchonella eyrensis*, gastropod *Euspira reflecta* and distinctive bivalves *Nototrigonia* and *Maccoyella barklyi*. The latter is one of the most ubiquitous molluscs in the Bulldog Shale.

almost universally endemic at genus and species level, incorporating unique bivalves such as *Cyrenopsis, Fissilunula, Pseudavicula, Tatella* and *Tancretella*, and the distinctive austral belemnites *Peratobelus* and *Tetrabelus*. This implies rapid speciation, perhaps in response to geographic isolation and the onset of cooler climatic conditions, once immigrant populations became established in the Australian epicontinental seaway.

Albian macroinvertebrate assemblages (from units such as the Toolebuc Formation and Allaru Mudstone) display lower diversity (~43 taxa) compared to those from the Aptian. Benthic molluscs in particular are restricted to only a few abundant but widely distributed forms, such as the Cretaceous bivalve *Inoceramus*. Faunal endemism is also generally lower than during the Aptian. For example, bivalves taxa such as *Anopaea* represent taxa that are widespread in both southern and northern high-latitude regions, and both the crustaceans (e.g. *Hoploparia* and *Torynomma*) and echinoderms (where diagnostic) comprise mainly cosmopolitan forms. The high proportion of epifaunal bivalves (e.g. *Inoceramus, Aucellina*) in the upper Albian deposits of the Toolebuc Formation and Allaru Mudstone is thought to reflect poorly oxygenated sea-floor sediments.

Albian pelagic macroinvertebrates include ammonites, nautiloids, belemnites and vampyromorph squid. Ammonites are particularly diverse and incorporate a mixture of southern endemics (e.g. the labeceratids *Labeceras* and *Myloceras*) and cosmopolitan forms (*Anisoceras, Beudanticeras, Hamites* and *Scaphites*). The latter are most dominant in Australian continental margin marine environments (e.g. the Bathurst Island Group of the Money Shoals Platform, Northern Territory) and appear to derive from contemporary assemblages in Europe

Albian (Lower Cretaceous) marine invertebrates. (Clockwise from top left) The giant bivalve *Inoceramus sutherlandi* and various crustaceans such as the shrimp *Hoploparia mesembria*, possible school shrimp *Mecorchirus* and crab *Torynomma*. All specimens are from the Toolebuc Formation/Allaru Mudstone series of central Queensland.

Albian (Lower Cretaceous) cephalopods. (Clockwise from top left) The ammonites *Labeceras*, *Myloceras* (Oodnadatta Formation, South Australia) and *Beudanticeras flindersi* (Toolebuc Formation, Queensland). Remains of the belemnite *Dimitobelus* (Oodnadatta Formation, South Australia), with positions of the solid calcite guard (bottom right) and phragmacone (bottom left) indicated in the interpretive diagram. Internal shell (up to 1 m long) of the giant vampyromorph squid *Trachyteuthis willisi* (Toolebuc Formation, Queensland). Shell of the nautiloid *Kummelonautilus hendersoni* (Oodnadatta Formation, South Australia).

and elsewhere in the northern hemisphere. The nautiloids *Cimomia* and *Kummelonautilus* and vampyromorphs *Boreopeltis* and *Trachyteuthis* also reflect northern origins. Conversely, the endemic ammonite and belemnite assemblages from the Australian inland seaway seem to have developed in response to progressive isolation of the epicontinental basin. The labeceratid ammonites (*Labeceras*, *Myloceras*) and dimitobelid belemnites (*Dimitobelus*) successfully radiated from this evolutionary centre into adjacent Gondwanan coastal shelf environments during the Late Cretaceous.

Sharks and bony fish

Considerable work has been undertaken on shark and bony fish faunas from the Lower Cretaceous deposits of Australia, in particular the upper Albian Toolebuc Formation and Allaru Mudstone of the northern Eromanga Basin in central and western Queensland. Rare chondrichthyan (sharks, rays and skates) and osteichthyan (bony) fish remains are also known from Aptian deposits such as the Bulldog Shale, which has produced the earliest known occurrence of the chimaerid (rat fish) *Edaphodon* from deposits near Lake Eyre, and isolated bones (mainly vertebrae) of small teleost fish from the opal mines at Coober Pedy and Andamooka. Large lamniform (the group that includes makos, tiger sharks etc.) shark vertebrae up to 125 mm in diameter have been recovered from the upper Aptian Darwin Formation near Darwin in the Northern Territory. These specimens are huge compared to the size of modern white pointers, suggesting that sharks >6 m in length were top predators in these continental margin environments.

Albian fish assemblages have been documented from the Toolebuc Formation and Allaru Mudstone of Queensland and from the upper Albian Gearle Siltstone of Western Australia. Sporadic middle Albian occurrences are known from the Coorikiana Sandstone of South Australia. Current work on this material (represented by isolated teeth, vertebral centra and scale accumulations) indicates mainly cosmopolitan affinities with taxa including:

- the hexanchoid (cow shark) *Notorynchus aptiensis*, particularly common in the continental margin sediments of the Gearle Siltstone;
- the wide-ranging *Paraorthacodus* known elsewhere from Late Cretaceous of Europe;
- the mitsukurinid-like *Anomotodon* or *Scapanorhynchus*;
- the large lamnids *Cardabiadon*, *Cretalamna appendiculata*, *Carcharias striatula*, *Paraisurus*, *Archaeolamna*, *Leptostyrax* and possibly *Cretoxyrhina*;
- the anacoracids *Microcorax*, *Squalicorax primaevus* and *Pseudocorax australis*, the last

Toothplate of the Lower Cretaceous chimaerid *Edaphodon eyrensis* from the Bulldog Shale deposits near Lake Eyre in northern South Australia.

Lower Cretaceous shark remains. (Left) Isolated tooth of the 3 m *Cardabiadon* found in the Albian Toolebuc Formation near Boulia, Queensland. (Right) Giant lamniform shark vertebra (125 mm diameter) from the Aptian Darwin Formation. (Bottom) Articulated vertebral column of an unidentified possible anacoracid shark from the Toolebuc Formation of Yamborra Creek near Maxwelton.

being the commonest shark genus in the Albian epicontinental deposits of Queensland.

Toothplates and fin spines of chimaeroids from the Toolebuc Formation correspond to at least one endemic genus and species *Ptykoptychion tayyo*. Significantly, isolated rostral teeth of the sawshark *Pristiophorus tumidens* are also present in the Toolebuc Formation and represent both the oldest known and the only southern hemisphere occurrence of this widespread taxon from the Cretaceous.

Albian bony fish remains are extremely abundant in the Toolebuc Formation. Much of this material comprises isolated teeth and bones; however, large articulated skeletons are also common. Described taxa incorporate several primitive forms including the armoured 1.5 m aspidorhynchid *Richmondichthys sweeti* (a toothless filter-feeding fish) and the predatory swordfish-like pachycormid *Australopachycormus hurleyi*. A number of endemic teleosts are also present:

- the elopiform *Flindersichthys denmeadi*, the predatory ichthyodectid *Cooyoo australis*, a large fish with skull length up to 400 mm, and '*Cladocyclus*' *sweeti*;
- the pachyrhizodontid *Pachyrhizodus marathonensis*, close relatives of which are well-known from the northern hemisphere;
- the small neoteleost *Dugaldia emmilta*.

These fish have important palaeobiogeographical links to South America, North America and Europe.

Lower Cretaceous bony fish remains from the Albian Toolebuc Formation of central Queensland. (Clockwise from top left) The skull (outlined) of the filter-feeding aspidorhynchid *Richmondichthys sweeti*, the skull and opercular region of *Pachyrhizodus marathonensis*, an acid-etched skull of *Cooyoo australis* and an incomplete skull of the bizarre swordfish-like predator *Australopachycormus hurleyi*.

Sea turtles

Australian Lower Cretaceous sea turtles (Chelonioidea) are confidently known only from the upper Albian Toolebuc Formation and Allaru Mudstone of Queensland, where they are among the commonest marine reptile fossils. Three distinct types have been named; all appear to be members of the widespread Cretaceous family Protostegidae. The first described was *Notochelone costata*, established on a partial carapace, plastron and limb girdle elements (from an unknown locality) by the famous English palaeontologist Sir Richard Owen in 1882. *Notochelone* is characterised by its small body size (<1 m) and low keel along the midline of the carapace.

A slightly larger sea turtle, *Bouliachelys suteri*, has been recorded from Toolebuc Formation deposits around Boulia in western Queensland. Represented by several skulls and some complete shells with associated limb elements, *Bouliachelys* is a primitive chelonioid close to the base of the protostegid radiation. Its robust skull has prominent crushing ridges on the palate and a serrated hooked beak, which suggests a diet of hard-shelled prey. Gut contents found within several fossil turtles from Boulia

Ichthyodectids such as *Cooyoo australis* from the Albian (Lower Cretaceous) of Australia often reached huge sizes. This skull and articulated anterior vertebral column found near Boulia in Queensland is nearly 500 mm in length.

An articulated carapace attributed to the small-bodied Albian (Lower Cretaceous) sea turtle *Notochelone costata*. This specimen is from the Toolebuc Formation deposits around Julia Creek in central Queensland.

Reconstruction of the Australian Early Cretaceous sea turtle *Bouliachelys suteri*. This species is known from isolated skulls, but associated postcranial remains indicate that a thin midline keel was present on the carapace.

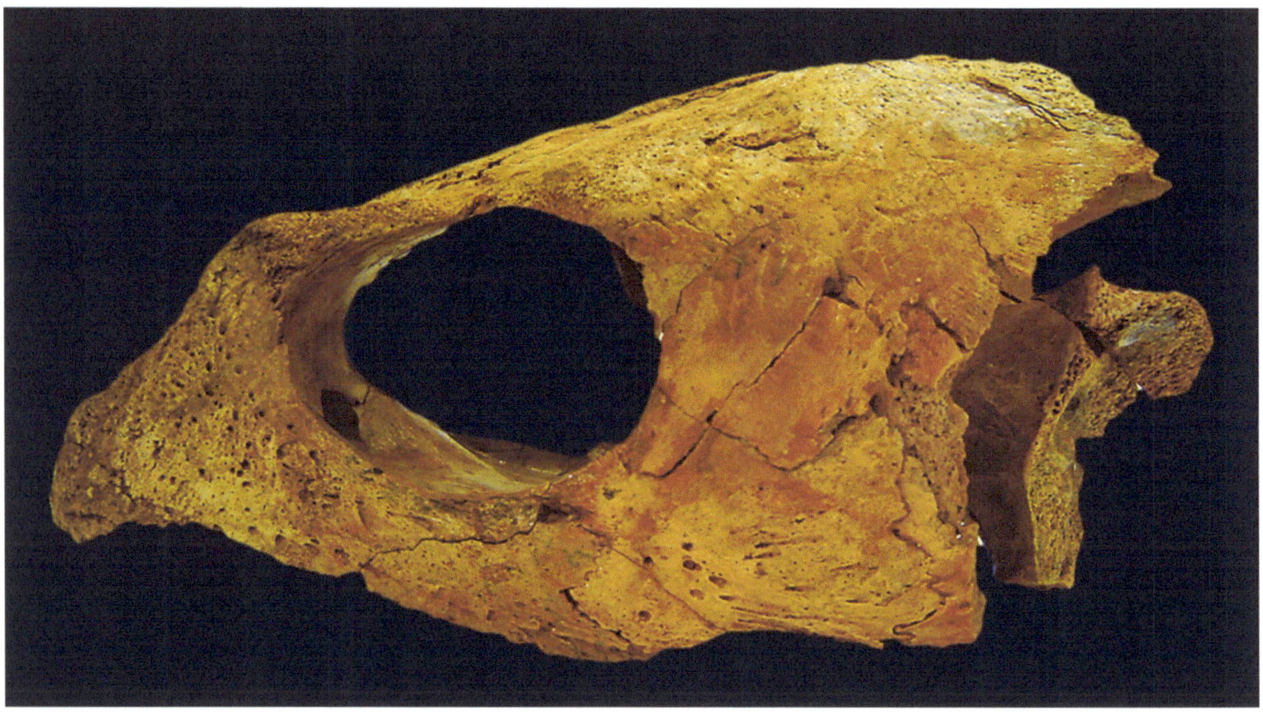

A superbly preserved skull of the Albian (Lower Cretaceous) sea turtle *Bouliachelys suteri* from Boulia, Queensland. There is a prominent nasal boss just in front of the orbit.

Limb bones of sea turtles from the Lower Cretaceous of Queensland. (Left) The 70 mm humerus of a small chelonioid, probably *Notochelone*. (Right) The 450 mm proximal humeral head of the giant *Cratochelone berneyi*.

consist almost entirely of chewed-up shell fragments from giant inoceramid bivalves.

The rarest and largest Australian Early Cretaceous sea turtle is *Cratochelone berneyi*, named on the basis of an associated fragmentary humerus, pectoral limb girdle and plastron from Sylvania Station, near Hughenden in central-northern Queensland. Comparison of *Cratochelone* bones with those of other large-bodied sea turtles from the Late Cretaceous of North America suggests a body length of ~4 m. Highly vascular bone structure implies that *Cratochelone* probably had augmented metabolic levels and pelagic habits similar to modern leatherback turtles. *Cratochelone* is the only Australian fossil sea turtle known from a single specimen, perhaps suggesting that it was an open-ocean inhabitant that only occasionally strayed into the shallow-water environment of the Australian inland sea.

Ichthyosaurs

Ichthyosaurs were a highly successful group of Mesozoic marine reptiles, characterised by a superficially fish-like body shape with triangular stabilising dorsal fin and vertically oriented crescentic tail fin. They were high-speed pursuit predators and a dominant component in marine vertebrate assemblages throughout the Early Cretaceous. Ichthyosaur remains are very common in the Lower Cretaceous deposits of Australia, and have been recorded from the Hauterivian-Barremian Birdrong Sandstone of Western Australia, the Aptian Bulldog Shale of South Australia, the Aptian Darwin Formation of the Northern Territory and most prolifically from

Reconstruction of the Australian Early Cretaceous ichthyosaur *Platypterygius australis*. This species is documented from Albian sites in Queensland and may also be present in Aptian strata from South Australia and the Northern Territory.

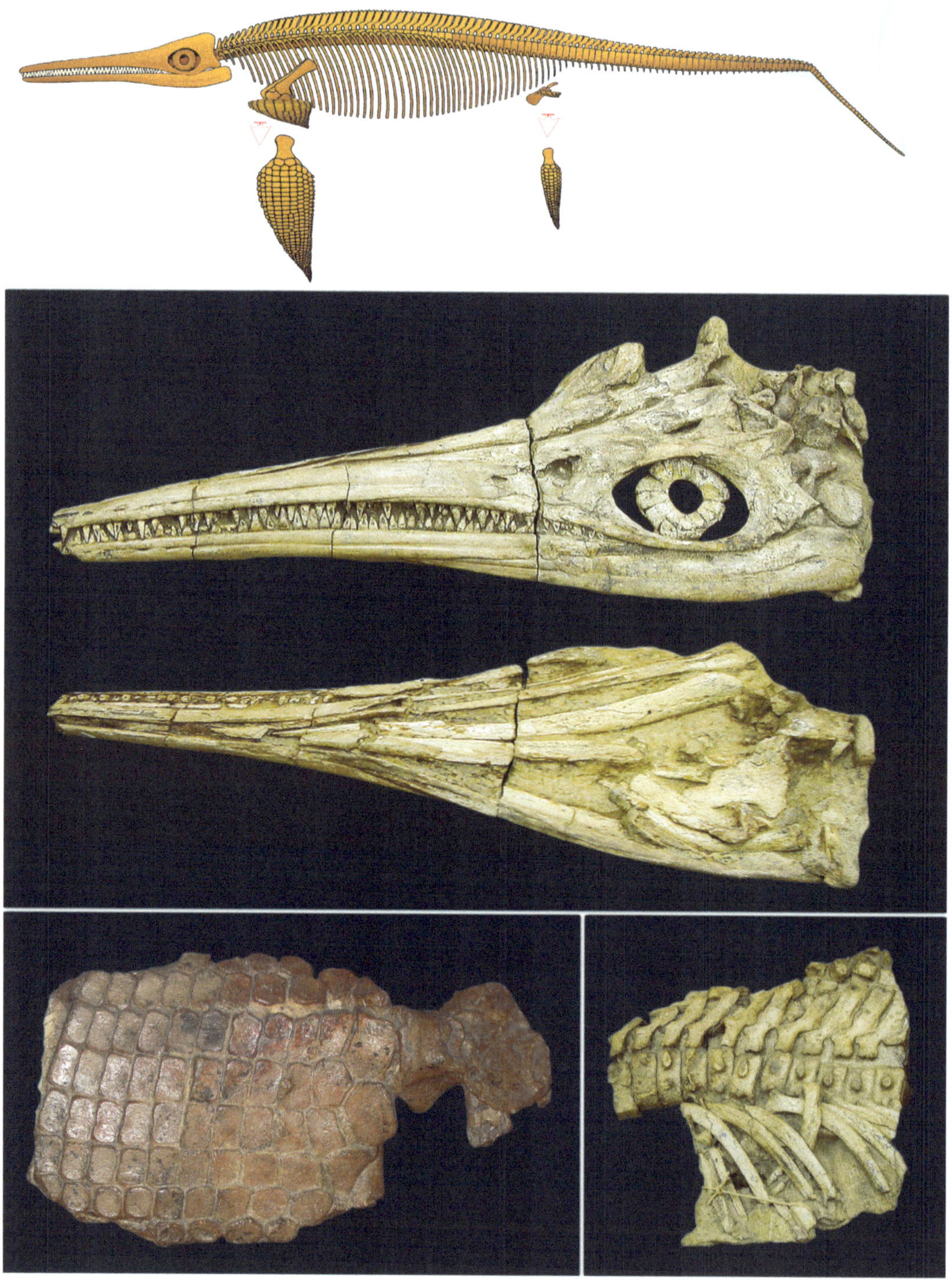

The Lower Cretaceous ichthyosaur *Platypterygius australis* from the Toolebuc Formation of central Queensland. (Clockwise from top) Skeletal reconstruction following Molnar (1991), skull of the famous Telemon Station specimen in dorsolateral and ventral views, an articulated anterior vertebral column and ribs, and the humerus and proximal forepaddle showing the tightly fitting block-like finger bones.

the upper Albian Toolebuc Formation and Allaru Mudstone (Rolling Downs Group) of Queensland. Much if not all of this fossil material can be assigned to the cosmopolitan ophthalmosaurid *Platypterygius*, the most widespread genus of Cretaceous ichthyosaur. The Australian species, *Platypterygius australis*, is represented by literally hundreds of skulls and skeletons. The very large pectoral fins are triangular in shape and comprise as many as nine rows of tightly fitting, rectangular finger bones. The powerful propulsive tail fin indicates shark-like swimming and a highly maneuvrable pursuit predator lifestyle. Preserved gut contents show that *Platypterygius australis* preyed on small fish, hatchling turtles and cephalopods, and that it scavenged bird carcasses.

The *Platypterygius* remains from around Australia are virtually indistinguishable. However, a fragmentary specimen from the Birdrong Sandstone of Western Australia displays slight differences in the structure of the humerus, that might suggest a separate species.

Plesiosaurs

Australian Lower Cretaceous plesiosaur remains are abundant and often exceptionally well-preserved. Representatives of five families have been recorded – Rhomaleosauridae, Pliosauridae, Aristonectidae, Elasmosauridae and Polycotylidae. The most complete examples of these groups come from the Aptian-Albian deposits of the Eromanga Basin in central and eastern Australia. Older occurrences have been reported from sediments in the Carnarvon Basin in north-western Western Australia. An isolated cervical vertebra recovered from Berriasian fluviatile deposits of the Barrow Group near Exmouth in Western Australia represents a small rhomaleosaurid-like plesiosaur of uncertain affinity. Rhomaleosaurids have also been reported from Hauterivian-Barremian rocks of the Birdrong Sandstone near Kalbarri in Western Australia.

Rhomaleosaurids were a group of primitive long-necked pliosauroids that dominated many Australian near-shore marine habitats throughout the Hauterivian-Aptian. Their remains were traditionally assigned to the genus *Leptocleidus* – an apparently globally distributed small-bodied (~2 m) form found in the Lower Cretaceous of England, South Africa and Australia. However, recent studies have shown *Leptocleidus* to be an artifical 'waste-basket' taxon that probably contains several closely related but distinct genera. The Birdrong Sandstone '*Leptocleidus*' material (consisting of several fragmentary skeletons) is therefore only provisionally retained in this genus as a separate species, '*Leptocleidus*' *clemai*, tenuously characterised on the basis of its larger size (~3 m).

Aptian rhomaleosaurids are best known from the Bulldog Shale of Coober Pedy and Andamooka in South Australia. Several opalised skeletons (including one with a near-complete skull, dubbed 'Eric' by the media) have been assigned to *Umoonasaurus demoscyllus*, a small-bodied (2.5 m) plesiosaur

Reconstruction of the Australian Early Cretaceous plesiosaur *Umoonasaurus demoscyllus*. This unusually archaic rhomaleosaurid thrived in the freezing high-latitude seas of South Australia 120 million years ago.

Reconstruction of a coastal area near Boulia 100 million years ago. Several *Platypterygius australis* feed on a group of recently hatched sea turtles. Gut contents from Australian *Platypterygius* specimens suggest a diverse diet of predominantly small prey such as belemnites, fish and young marine reptiles. We do not know if ichthyosaurs lived and hunted in groups, but healed tooth-marks on bones suggests that *Platypterygius* individuals may have fought each other over food and mates.

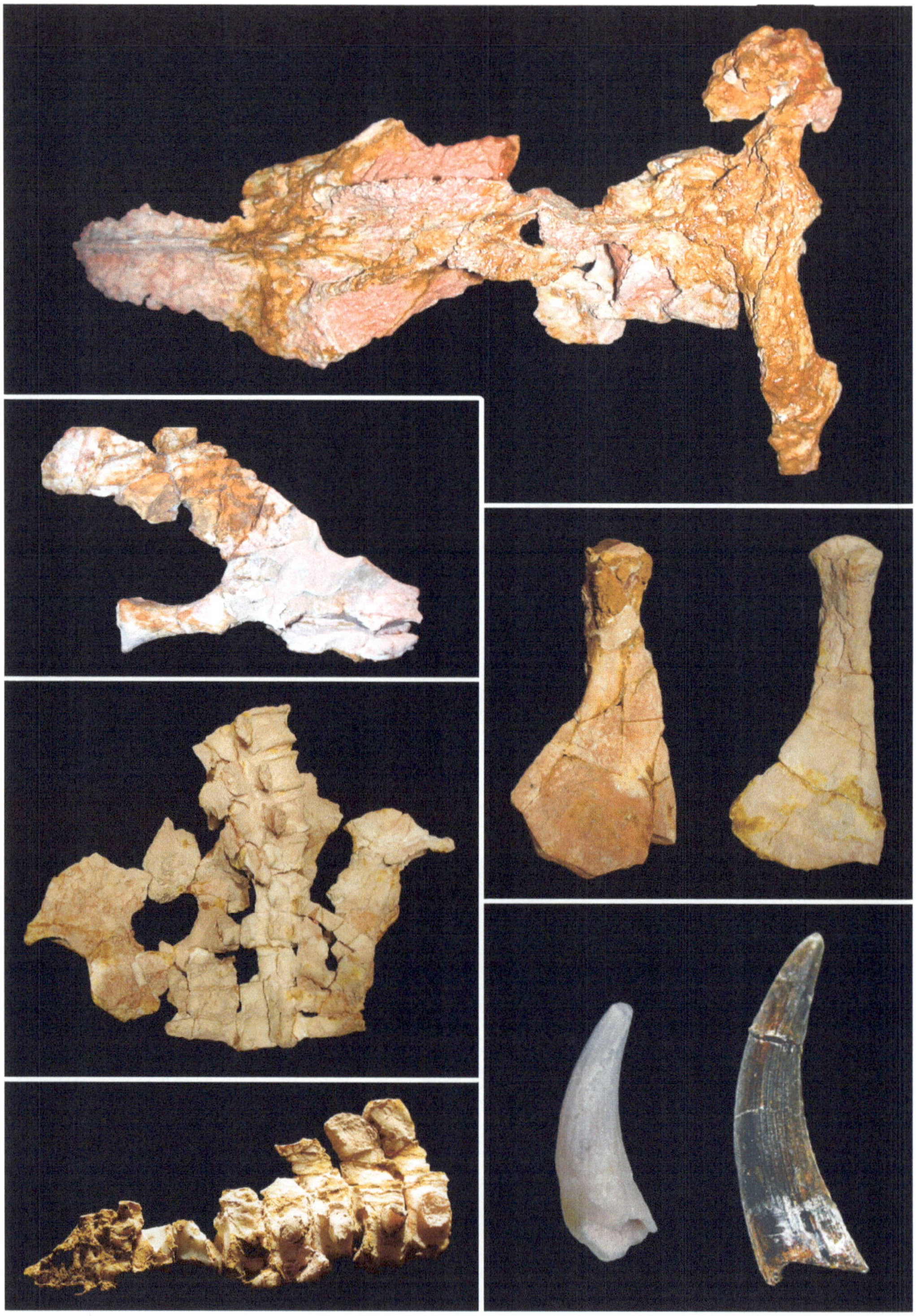

Skeletal elements of Lower Cretaceous rhomaleosaurid-like plesiosaurs. (Clockwise from top) Skull (dorsal view) of 'Eric', the famous opalised skeleton of *Umoonasaurus demoscyllus* from the Bulldog Shale of Coober Pedy; femur (left) and humerus (right) of *Umoonasaurus demoscyllus*; isolated teeth from non-marine deposits of the Griman Creek Formation (left) and Wonthaggi Formation (right); articulated tail section of *Umoonasaurus demoscyllus* showing fused tail tip; pelvis of *Umoonasaurus demoscyllus* comprising a partial pubis, ischium and vertebral column; and the scapula of *Umoonasaurus demoscyllus* in lateral view. Comparable views are shown in Kear (2006) and Kear *et al.* (2006).

Reconstruction of the giant pliosaurid *Kronosaurus queenslandicus* from the Aptian-Albian of Australia. *Kronosaurus* remains are widespread throughout deposits in the Eromanga Basin.

that uniquely possessed thin crests along the midline of the skull and above the orbits. *Umoonasaurus* is surprisingly archaic despite its relatively late age (Aptian), and simultaneously represents the most primitive and the last surviving rhomaleosaurid. The apparent restriction of *Umoonasaurus* to the near-freezing high-latitude (70°S) inland seaway of Early Cretaceous Australia suggests that it might have possessed adaptations (e.g. heightened metabolic levels) to cope with low water temperatures. Survival of ancient lineages such as *Umoonasaurus* is a common phenomenon in the Australian Cretaceous and might be a product of isolation in low-temperature high-latitude regions.

Isolated rhomaleosaurid-like remains, mainly isolated teeth, have been described from Aptian-Albian non-marine strata of the Wonthaggi and Eumeralla formations in Victoria, from the Griman Creek Formation of Lightning Ridge in New South Wales and from the Surat district of Queensland.

Snout (dorsal view) of *Kronosaurus* from the Albian Toolebuc Formation of Queensland. This 1.5 m section of skull belongs to a juvenile *Kronosaurus* skeleton nearly 9.5 m in total length.

Snout and lower jaw of a *Kronosaurus* specimen from the Lower Cretaceous of Queensland. The massive interlocking teeth are over 150 mm in length and would have killed and dismembered prey.

Interestingly, these deposits represent cold high-latitude rivers and thus provide further evidence of plesiosaurs living in low-temperature inland waterways near the Early Cretaceous South Pole.

Large-bodied predatory pliosaurids are another distinctive element of Australian Aptian plesiosaur faunas. Numerous fragmentary specimens have been recovered from the Wallumbilla Formation deposits in Queensland and New South Wales and from the Bulldog Shale of South Australia. Large pliosaurid remains also frequently occur in the upper Albian Toolebuc Formation/Allaru Mudstone of Queensland. This material is currently assigned to a single genus, *Kronosaurus*, with the Aptian remains representing a distinct species, *Kronosaurus queenslandicus*. *Kronosaurus* was a gigantic pliosaurid, possibly >10 m in length. Isolated cranial remains indicate skulls of over 2.5 m, and conical coarsely striated teeth up to 150 mm high. The characteristically flattened skull and upward facing orbits might indicate that *Kronosaurus* was an ambush predator, attacking prey by surprise from below. Gut contents preserved in some *Kronosaurus* specimens show a diet of marine reptiles (elasmosaurid plesiosaurs) and large sharks.

Aristonectids are a rare plesiosaur group in Australia, reported only from the Aptian Bulldog Shale deposits of Andamooka and Coober Pedy in South Australia. Aristonectids were long-necked plesiosauroids that have been variously allied with elasmosaurids and Jurassic cryptoclidoids. Aristonectids are known elsewhere from the Upper Jurassic of Europe (England) and North America and from the Upper Cretaceous high-latitude sediments of Antarctica, New Zealand and Patagonia. It has been suggested that Late Cretaceous aristonectids might have been specialised for low-temperature environments. The only Australian aristonectid is *Opallionectes andamookaensis*, named from a single opalised skeleton (lacking the skull), with small needle-like teeth that were possibly used for filter-feeding. *Opallionectes* displays a mixture of derived and primitive traits that place it as a structural and stratigraphical intermediate between Late Jurassic aristonectids from the northern hemisphere, and

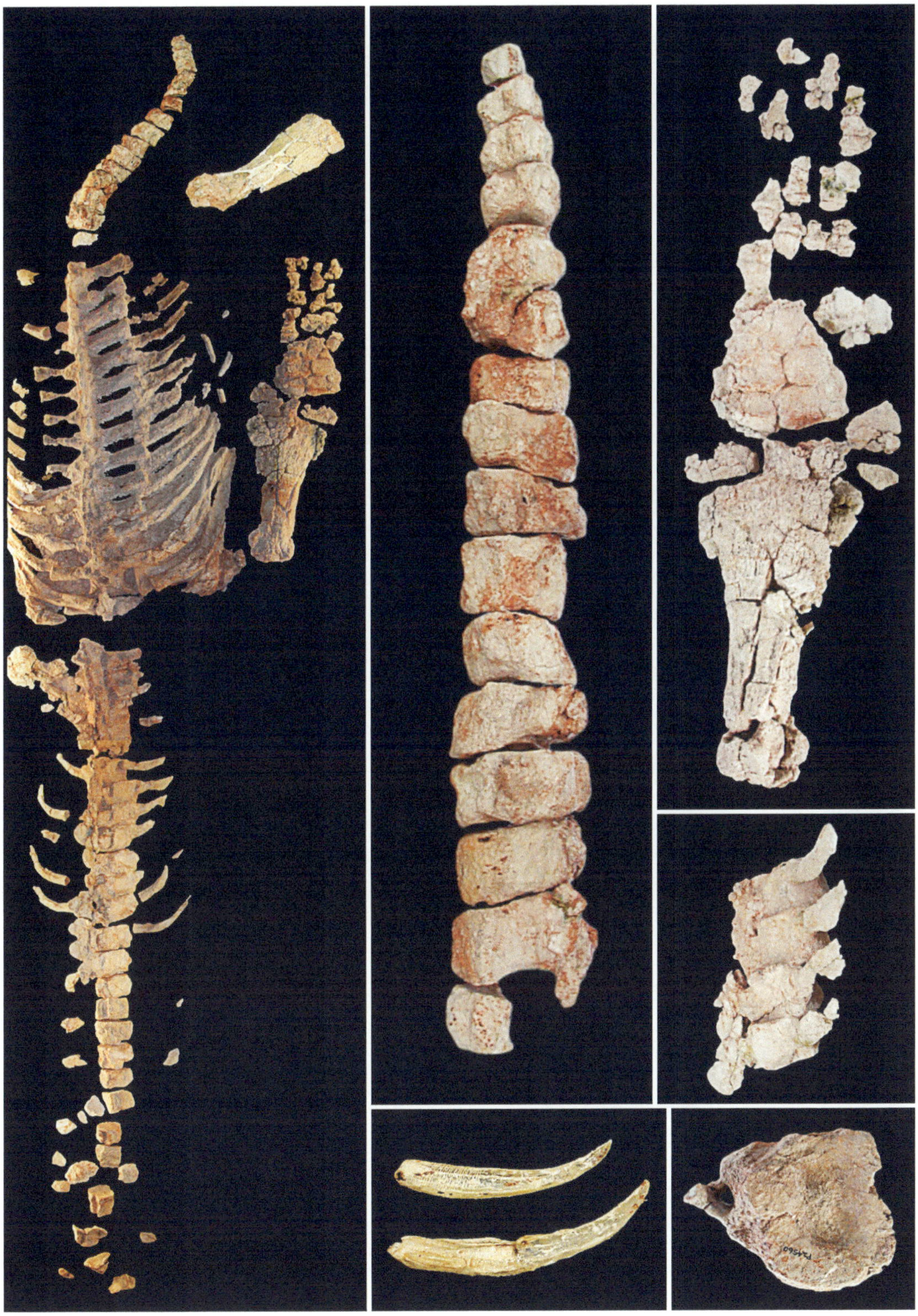

Skeletal elements of the aristonectid *Opallionectes andamookaensis* from the opal mines of Andamooka in South Australia. (Clockwise from left) Articulated skeleton, caudal centra series, forelimb, cervical vertebrae in side and anterior view of the articular surface, and teeth. Comparable views of the holotype are shown in Kear (2006).

Reconstruction of the elasmosaurid *Eromangasaurus australis* from the Albian Toolebuc Formation of Queensland. Elasmosaurids are widespread in the Cretaceous deposits of Australia.

advanced Late Cretaceous taxa from the southern polar regions.

During the Albian, Australian plesiosaur faunas appear to have undergone a distinct reorganisation with elasmosaurids (very long-necked Cretaceous plesiosaurs) replacing rhomaleosaurids as the most ubiquitous group. Australian elasmosaurids are geographically and stratigraphically widespread with remains found in the Aptian to Albian of Queensland (Rolling Downs Group), New South Wales (Wallumbilla Formation), South Australia (Bulldog Shale and overlying Albian Oodnadatta Formation) and the Northern Territory (Darwin Formation). Most of this material is indeterminable except for the nearly complete skull of *Eromangasaurus australis*, from the upper Albian Toolebuc Formation near Maxwelton in central-northern Queensland. *Eromangasaurus* displays several distinctive features that indicate a possible close relationship with elasmosaurs from the Americas. The holotype skull of *Eromangasaurus* is also of interest because it shows evidence of having been bitten and crushed by a large predator, presumably a *Kronosaurus* based on the size and spacing of the huge toothmarks.

Polycotylids are a rare component of the Australian Aptian-Albian plesiosaur fauna. Their occurrence in the Early Cretaceous of Australia is surprising, given that these distinctive long-snouted plesiosauroids are primarily a Late Cretaceous northern hemisphere group (Cenomanian to Maastrichtian of North America, South America, Europe, North Africa and Japan). The Australian remains constitute the oldest known examples of the family, with a partial opalised skeleton recovered from the Wallumbilla Formation at White Cliffs in New South Wales. Very well-preserved polycotylid fossils have been discovered in the upper Albian Toolebuc Formation of central Queensland. These represent a new, as yet unnamed, primitive polycotylid closely related to Late Cretaceous taxa from the northern hemisphere.

Elasmosaurid remains from the Lower Cretaceous of Queensland. (Clockwise from top) The skull of *Eromangasaurus australis* in side and ventral views and its cervical vertebrae, and the articulated humerus and proximal paddle of an indeterminate elasmosaurid from the Aptian Wallumbilla Formation.

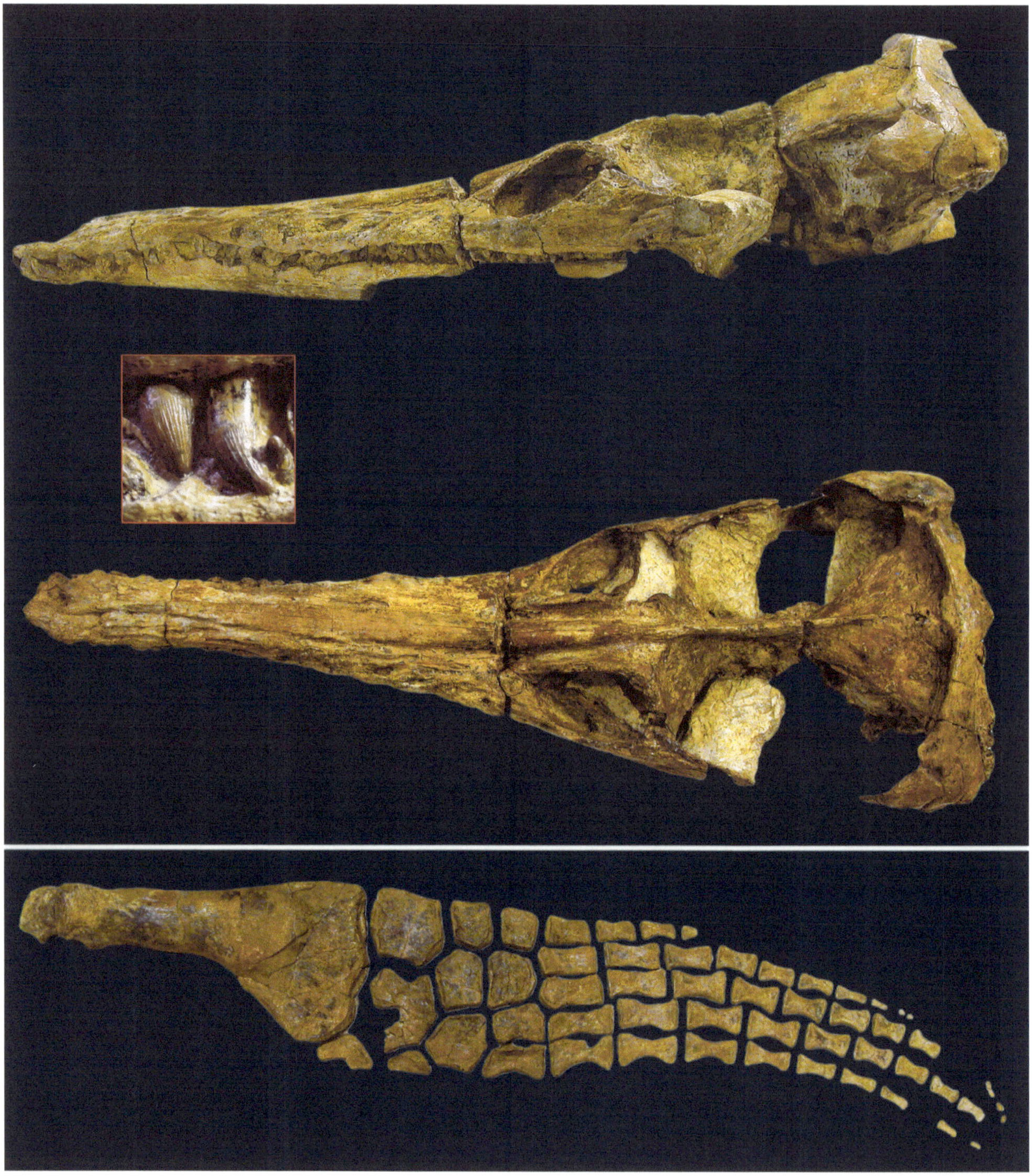

Remains of an unamed polycotylid, popularly dubbed the Richmond pliosaur, from the Albian Toolebuc Formation of Queensland. (Top) Skull (with teeth magnified) in side and dorsal views. (Bottom) Articulated forelimb.

6

LIFE ON THE SHORE: EARLY CRETACEOUS NON-MARINE ANIMALS AND PLANTS

The Australian continent was geographically isolated at the beginning of the Cretaceous (Berriasian) but maintained a continuous landsurface with India along its western margin, with the Queensland Plateau and Lord Howe Rise constituting an eastern coastal fringe. River systems active in the Canning, Carpentaria, Eromanga and Surat basins drained towards the north-western and north-eastern coastlines. Rivers also flowed through the southern Perth Basin of Western Australia prior to separation from India. The southern rift valley that had formed between Australia and Antarctica during the Jurassic supported a complex series of rivers and lakes. These deposited vast amounts of sediment into the modern Otway, Gippsland and Bass basins.

Following final separation from India in the Valanginian (140 million years ago), much of the western continental margin became inundated by sea. However, terrestrial environments persisted until the Barremian (130 million years ago), when a major rise in global sea-levels flooded large areas of north-eastern Australia (Carpentaria and Eromanga basins). This geographically isolated land-based organisms and produced a high proportion of endemic and relictual animals and plants. Continued sea-level rises during the Aptian further restricted terrestrial habitats in central and Western Australia, the Queensland Plateau and the south-eastern rift valley, which partially flooded to the west but remained largely landlocked. Upland areas in north-eastern Australia (currently forming the Great Dividing Range), developed via extensive crustal folding and volcanic activity in the late Palaeozoic (Permian), now provided new coastal environments. These expanded with the regression of the inland sea in the late Albian (around 100 million years ago), and were followed by brackish lagoons and estuaries in the Surat Basin. Further to the east, coal-forming swamps (today represented by the Styx River Coal Measures) and rivers drained eastwards off the Queensland Plateau's outboard coastal margin.

Some of Australia's oldest Cretaceous terrestrial fossils are derived from the Broome Sandstone, which crops out near Broome in far north-western Western Australia (Canning Basin). The Broome Sandstone is exposed on wave-cut platforms that reveal spectacular dinosaur trackways. Lower Cretaceous fossil assemblages are also known from the Barremian-Aptian Strzelecki Group including the Koonwarra Fossil Beds near Leongatha and the Flat Rocks Site near Inverloch in southern Victoria. These have yielded evidence of diverse faunas and floras that lived in the ancient rift valley stretching between Australia and Antarctica ~120 million years ago.

Australia's record of terrestrial animals and plants from the Albian includes remains from near-shore marine deposits in Queensland, where carcasses were washed out to sea. The Toolebuc Formation, Allaru Mudstone and Mackunda Formation (Eromanga Basin) have all produced rare but superbly preserved specimens of dinosaurs,

Eagles Nest is a classic locality for Australian Lower Cretaceous dinosaurs. In 1903 the geologist William Hamilton Ferguson discovered the first Australian dinosaur fossil – a claw belonging to a carnivorous theropod, that was sent to the British Museum in London for identification. Systematic exploration of Lower Cretaceous deposits along the Victorian coast have revealed a diverse non-marine fauna comprising dinosaurs, bony fish, turtles, relict temnospondyl amphibians and tiny mammals. Image produced by Kay Ballard-Hamilton, courtesy Thomas Rich.

One of Australia's most productive Aptian non-marine fossils sites is Flat Rocks in Victoria. This locality has yielded a wealth of dinosaur and early mammal remains. Image produced by Lesley Kool, courtesy Thomas Rich.

pterosaurs and birds. More fragmentary but nonetheless intriguing fossil material has been derived from the opal mines at Lightning Ridge in New South Wales (Griman Creek Formation) and from coastal cliffs in the Otway Ranges (Otway Group) of southern Victoria. Both these regions were dominated by river systems ~110 million years ago.

Significant Australian Lower Cretaceous non-marine deposits and localities

The Wonthaggi Formation/Koonwarra Fossil Beds (Strzelecki Group), together with the stratigraphically younger Eumeralla Formation (Otway Group), comprise a series of finely laminated sandstones and mudstones that crop out within the coastal Gippsland and Otway basins of south-eastern Australia. These units were laid down in a mid-Cretaceous floodplain formed by rifting between Australia and Antarctica. The sediments are typically volcanic in origin and probably derived from a tectonically active region that was associated with the formation of the Tasman Sea.

The Wonthaggi and Eumeralla formations record a progressive change in depositional setting from predominantly meandering to braided river systems. The older Wonthaggi Formation is characterised by channel deposits with thick accumulations of cross-bedded sediments that were rapidly accumulated by flash floods or seasonal snowmelt flows. In contrast, the younger Eumeralla Formation is dominated by lenses of fine sand derived from successive river channels. The Koonwarra Fossil Beds differ from both these environments in representing a low-energy lake deposit dominated by very fine-grained clays.

Age estimates for the Wonthaggi Formation/Koonwarra Fossil Beds and Eumeralla Formation have been determined from fission-track dating and microfossil data. The most recent work suggests a middle Valanginian to Aptian age for the Wonthaggi Formation/Koonwarra Fossil Beds (the latter considered Barremian-Aptian only), with the Eumeralla Formation being upper Aptian-lower Albian.

The opal mines of Lightning Ridge in northern New South Wales are a major source of Albian non-marine fossils. Image courtesy Henk Godthelp.

Many opal fossils are recovered as a by-product of mining. Opal-bearing rock is tumbled with water in a cement mixer (shown here at Lightning Ridge, New South Wales) to break down the soft claystone. The opal is extracted by hand as it exits the tumbler along a conveyer belt. Opal fossils are compositionally identical to the opal gemstone. Indeed, many pieces of jewellery opal are cut from the fossilised remains of ancient animals and plants. Opal fossils are often found in tailings heaps, as seen here. Image courtesy Henk Godthelp.

The Griman Creek Formation (Rolling Downs Group) crops out in the Surat Basin of northern New South Wales and south-eastern Queensland. The unit comprises interbedded sandstones, siltstones and mudstones that are rich in non-marine mollusc fossils (hyriid bivalves, thiiarid and viviparid gastropods) and plant root impressions. They preserve sediment cross-bedding that is suggestive of a freshwater to brackish-estuarine origin. Some horizons of the Griman Creek Formation are bleached white by weathering. This is best seen in the areas around Lightning Ridge, where the clay-rich sediments contain opal. The opal at Lightning Ridge occurs as concretionary structures ('nobbies'), or as replacements of shells, wood or bone. The process of opal formation at Lightning Ridge is poorly understood, although it has been linked to bacterial precipitation of silica from acidic groundwater during the Cretaceous.

Rich microfossil assemblages have been used to determine the age of the Griman Creek Formation as lower-middle Albian. As most microfossil material has been leached from the opal-bearing strata at Lightning Ridge, it is difficult to establish precise dating for this locality.

Estimates of paleolatitude place south-eastern Australia near the southern polar circle throughout much of the Jurassic-Cretaceous. The Wonthaggi Formation/Koonwarra Fossil Beds and Eumeralla Formation were laid down at about 60°–80°S (specific estimates of 66.8°S have been determined for the Otway Group, and 77.8°S for the Strzelecki Group sediments). They preserve sedimentary structures, stable isotopes and palaeofloras which indicate highly seasonal cool to very cold (possibly with

Mining for opal fossils can be dangerous, as shown by this excavation at Lightning Ridge. Many older underground mines have been 'ballroomed' – rock supports have been removed, and replaced with wooden props. Over time, drying and contraction of the soft claystone makes the mine roof unstable, resulting in collapse. Image courtesy Henk Godthelp.

winter freezing) climatic conditions. Sediments of the Griman Creek Formation are thought to have been deposited in an Early Cretaceous high-latitude zone (~60°S), but correspond to a warmer, temperate climatic period around the middle-late Albian.

Paleotemperatures for the southern Australian region averaged ~12°C. However, continental surface air temperatures for south-eastern Australia may have dropped close to 0°C. There is possible evidence of permafrost in Cretaceous sediments from Victoria and glacial tillites from South Australia.

Land plants

The Australian Lower Cretaceous land plant record has been examined in some detail using palynofloras (spores and pollen) and macrofossils. The vegetation is typified by continent-wide coniferous forests (podocarps and araucarians) associated with open heathlands and river bank/coastal communities. Berriasian plant macrofossils obtained from subsurface drillcores in the Otway Basin indicate an understorey of bennettitaleans (cycad-like plants). Specimens from the Berriasian-Valanginian Algebukina Sandstone, near Oodnadatta in South Australia (Eromanga Basin), similarly show that inland forests comprised canopy conifers with an understorey of seed-ferns (*Rienitsia variabilis*), pentoxylaceans (*Taeniopteris spatulata*), cycads and bennettitaleans. Some regional endemism is evident in Berriasian-Valanginian floras from the Otway and Gippsland basins, which seem to have supported a more densely forested but less diverse flora than those further north. For example, the Cretaceous floodplain communities of the Eromanga and Surat basins in north-eastern Queensland have yielded podocarp (*Elatocladus*) and araucarian conifers (*Brachyphyllum*), together with ginkgophytes (*Ginkgoites australis*) as canopy elements. Understorey plants include ferns (e.g. *Phyllopteroides*), pentoxylaceans (*Taeniopteris*), bennettitaleans and horsetails (*Equisites*). The

An opalised plesiosaur tooth, broken by an ancient stream flow 100 million years ago. Specimens such as this, photographed in an opal mine wall at Lightning Ridge, were accumulated by rivers dumping fine sediments into a coastal estuary system. Image courtesy Henk Godthelp.

earliest Cretaceous macrofloras from Western Australia are comparatively poorly preserved (fragmentary material is known from the Broome Sandstone). However, well-documented palynofloras indicate strong floral links with eastern Antarctica and India. In contrast, the plants from eastern Australia had closer ties with assemblages from the western Antarctic region.

During the Barremian-Aptian, rising sea-levels progressively restricted terrestrial floras. Rich macrofossil material from the Otway and Gippsland basins, in particular the Koonwarra Fossil Beds, indicates a cool, humid montane-adapted forest that included podocarps (*Bellarinea*, *Brachyphyllum*, *Podostrobus*), araucarians (*Araucaria*) and *Ginkgoites* as canopy elements. Driftwood from Aptian marine deposits (e.g. Bulldog Shale of South Australia) shows closely spaced growth rings indicative of cool climates and marked seasonality. Understorey vegetation was dominated by ferns and pentoxylaceans such as *Taeniopteris daintreei*, which probably grew in large open areas. Bennettitaleans and cycads were increasingly rare during the Aptian, with horsetails (*Equisetum*), ferns, mosses, liverworts and some early flowering plants (probably low-growing and herbaceous) occupying river and lake margins.

Around the Aptian-Albian boundary, a major shift in vegetative composition saw an increase in the distribution and diversity of flowering plants. Magnolias are especially well-documented from the Albian pollen record. They seem to have been more diverse in northern Australia, where angiospermous macrofossils occur in the swampy palaeoenvironments of the Styx and Burrum coal measures

Lower Cretaceous plant fossils from the Berriasian-Valanginian Algebuckina Sandstone of South Australia. (Clockwise from top left) A delicate foliage impression (with outline highlighted) of the conifer *Elatocladus planus*, foliage of the pentoxylacean *Taeniopteris spatulata*, trunk structure of the giant lycopsid moss *Nathorstianella babbagensis* and foliage of the cycad *Nilssonia*.

(north-eastern Queensland). Ferns (*Phyllites*), horsetails, pentoxylaceans (*Taeniopteris*, which has disappeared from contemporary floras in the south), cycads, araucarian conifers and *Ginkgoites* are also present in these deposits. During the Albian, the southern Austro-Antarctic rift valley saw the evolution of new floral elements, possibly in response to an expansion of floodplain environments and the onset of warmer climatic conditions. The osmundaceous fern (*Phyllopteroides*) and taxodiaceaean conifers (*Geinitzia*) appear in the record, and the modern ginkgophyte genus *Ginkgo* replaced *Ginkgoites* in the canopy assemblage. Late Albian coastal areas around the inland sea (Eromanga Basin) supported araucarian/podocarp forests (identified from wood in marine units such as the

Living araucarian conifers are virtually identical to their Mesozoic ancestors. Their broad-leaf foliage and reproductive cones are common fossils in Lower Cretaceous non-marine deposits throughout Australia.

Toolebuc Formation) and cycads (*Nilssonia macronatum*), with ferns, mosses and liverworts in heathland and boggy areas.

Non-marine invertebrates

The current record of Australian Lower Cretaceous non-marine macroinvertebrates is substantial but mainly limited to a few highly productive localities. Some remains have been recovered from near-shore marine rocks including the Valanginian Trinity Well Sandstone Member of the Cadna-owie Formation (South Australia) and the Aptian Wallumbilla Formation (White Cliffs, New South Wales). These have produced freshwater unionoid bivalves (*Alathyria coatsi, Hyridella whitecliffsensis*) and Australia's oldest known non-marine gastropod – the viviparid *Notopala alboscopularis. Notopala* is still common in freshwater–estuarine habitats today.

By far the richest Lower Cretaceous deposits for non-marine invertebrates are the Barremian-Aptian Koonwarra Fossil Beds of southern Victoria. Extensive sampling of these ancient lake sediments has produced more than 80 species, including:

- bivalves (*Alathyria coatsi* and *Mesohydridella ipsviciensis* are also known from other nearby localities);
- freshwater bryozoans;
- a possible earthworm;
- harvestman spiders;
- crustaceans (the syncarid *Koonaspides indistinctus*, a shrimp or anostracan, an ostracode and cladoceran);
- more than 70 species of insects.

The Koonwarra insects represent the best-documented Lower Cretaceous assemblage of this kind from the southern hemsphere. They incorporate a number of groups that are still extant. Lake-dwelling dragonfly, scorpionfly, caddisfly and mayfly larvae (*Australurus plexus*) dominate, together with

Lower Cretaceous plant remains. (Top left) Delicate ginkgophyte foliage *Ginkgoites* from the Aptian Koonwarra Fossil Beds of southern Victoria. (Top right) An opalised conifer cone from the Griman Creek Formation at Lightning Ridge, New South Wales. (Bottom) Foliage of the pentoxylacean *Taeniopteris daintreei* from the Aptian Koonwarra Fossil Beds of southern Victoria.

aquatic hydrophilid beetles and stream-inhabiting mayflies (*Promirara cephalota* and *Dulcimanna sculptor*) that probably washed in from nearby watercourses. Airborne (e.g. dragonflies, scorpion flies and parasitic wasps) and terrestrial (e.g beetles and pucilid fleas, suggesting the presence of host mammals) insects are also present, presumably blown or fallen onto the water surface. The Koonwarra assemblage is indicative of shallow cool-water conditions and is thought to have undergone periodic mass mortality, perhaps in response to winter freezing of the lake's surface.

Insect fossils are also known from the upper Albian marine Toolebuc Formation, which has

Opalised invertebrate fossils from the Lower Cretaceous Griman Creek Formation of New South Wales. (Left) The unionoid bivalve *Hyridella macmichaeli* in side (top and bottom) and (middle) commissural view. Note the vertebrate bite marks on the upper valve of the shell. (Top right) A 'yabby button' or gastrolith from a decapod crustacean. (Bottom right) A second beautifully opalised specimen of *Hyridella macmichaeli*. Image courtesy Robert Smith and the Australian Opal Centre.

Wing of the dragonfly *Aeschnidopsis flindersiensis* from the Albian Toolebuc Formation near Richmond in central Queensland.

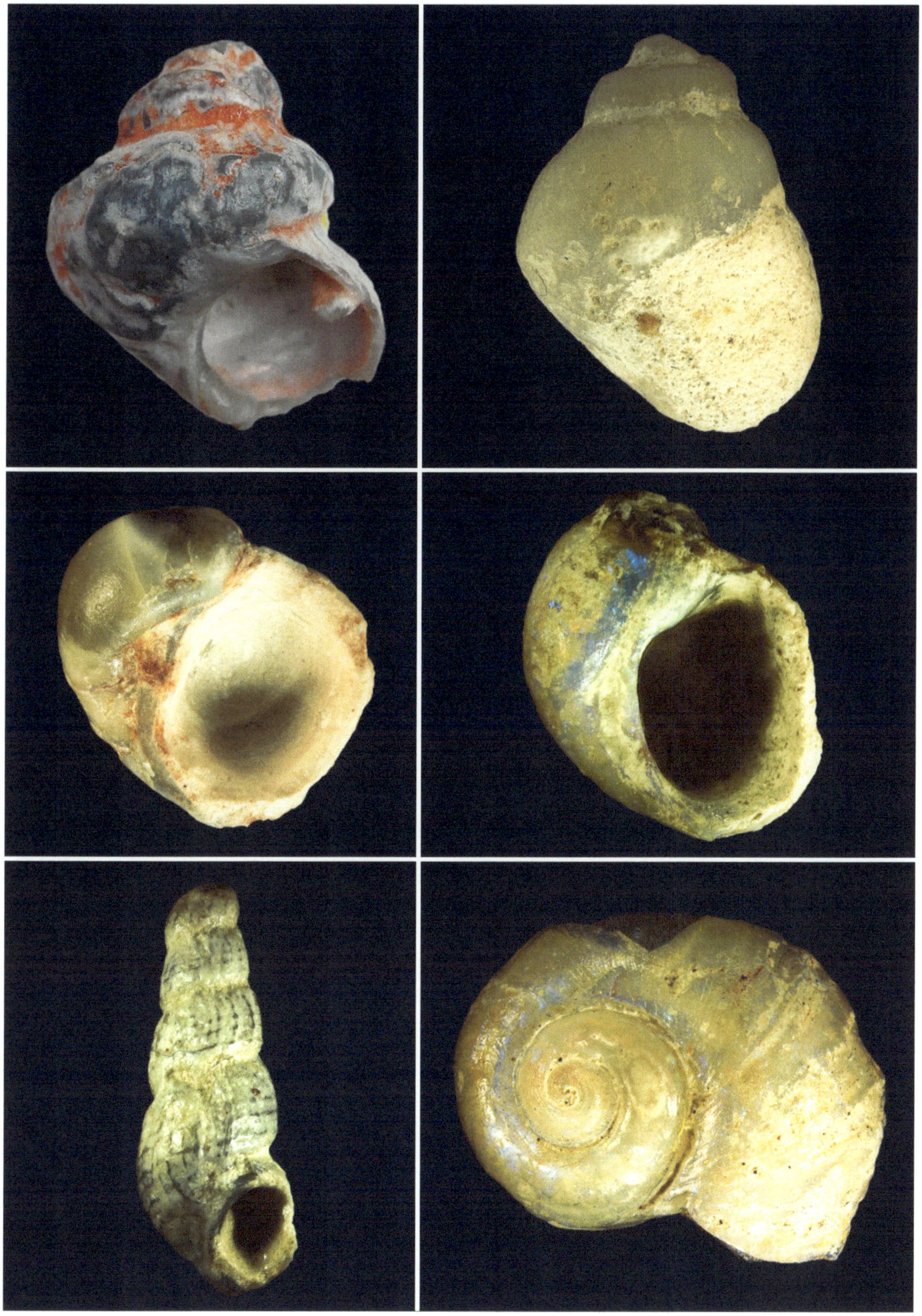

Opalised gastropods from the Albian (Lower Cretaceous) Griman Creek Formation of New South Wales. (Clockwise from top left) The viviparids *Albianopalin benkeari* and *Fretacaeles gautae,* possible succineid and camaenid land snails, the thiiarid *Melanoides godthelpi* and another possible pulmonate.

Specimens of the primitive teleost *Leptolepis koonwarri* from the Lower Cretaceous Koonwarra Fossil Beds in southern Victoria.

yielded a single wing impression of the dragonfly *Aeschnidopsis flindersiensis* and an unnamed mecopteroan or scorpion fly. The lower-middle Albian Griman Creek Formation at Lightning Ridge has produced decapod crustaceans (known only from gastroliths), and a number of non-marine unionoid bivalves and gastropods. The bivalves include *Alathyria jaqueti*, *Palaeohyridella godthelpi*, *Megalovirgus wintonensis* and *Hyridella macmichaeli*; these taxa are known throughout the Cretaceous in Australia. The gastropods comprise both freshwater viviparids (*Notopala*, *Albianopalin benkeari*, *A. lizsmithae*, *Fretacaeles gautae*) and brackish water thiarids (*Melanoides godthelpi*), suggesting a complex estuarine palaeohabitat. Terrestrial pulmonates (camaenids and possible succineids) are also present; they were probably washed into the aquatic deposits or fell from overhanging vegetation.

Non-marine sharks and bony fish

Australian Cretaceous non-marine shark remains are extremely rare. However, opalised lamniform teeth and the jaw plate of a small ray have been found in the Albian Griman Creek Formation at Lightning

Lower Cretaceous non-marine fish remains from the Griman Creek Formation of Lightning Ridge. (Clockwise from top left) Rare lamniform shark teeth and a bony scale from a marine aspidorphynchid (images courtesy Henk Godthelp), the skull of an unidentified bony fish that may represent an early eel (image courtesy Robert Smith) and a toothplate of the widespread lungfish *Metaceratodus woollastoni.*

Ridge in north-western New South Wales. The shark teeth are unidentified but are comparable to marine forms, and may represent individuals moving upstream from estuaries and coastal lagoons.

The Barremian-Aptian Koonwarra Fossil Beds and lower to middle Albian Griman Creek Formation have also yielded bony fish fossils. The Koonwarra remains are thought to have accumulated via seasonal winter kills, when large numbers of fish were suffocated as lakes froze over. Populations probably restocked from adjacent streams emptying into the lake system during the summer thaw.

The Koonwarra Fossil Beds have produced a primitive teleost, *Leptolepis koonwarri*, and a palaeoniscoid, *Coccolepis woodwardi*. Other taxa include an archaeomaenid *Wadeichthys oxyops* (also known from the Australian Jurassic), a clupeioform *Koonwarria manifrons* and an unidentified ceratodont lungfish. Ceratodonts, such as the large-bodied *Metaceratodus wollastoni*, have also been recorded from the Aptian Wallumbilla Formation of White Cliffs in New South Wales and possibly from the Bulldog Shale of South Australia. *Archaeoceratodus avus* is a ceratodont known from the Wonthaggi Formation at Cape Patterson in southern Victoria and from the Middle Triassic Hawkesbury Sandstone; this implies an exceptionally long evolutionary history stretching throughout much of the Mesozoic.

Teleosts (*Leptolepis crassicaudata*) and other enigmatic bony fish (*Psilichthys selwyni*) have been documented from the Albian Otway Group of southern Victoria. Teleost-like jaw fragments and vertebrae are known from the Griman Creek Formation at Lightning Ridge, together with scales of the marine aspidorhynchid *Richmondichthys*, and a small rather structurally advanced skull that might be an early eel. The ceratodonts at Lightning Ridge include the ubiquitous *Metaceratodus wollastoni* (reported elsewhere in the Albian Mackunda Formation, Albian-Cenomanian Winton Formation and Palaeogene-Neogene deposits of central Australia, Queensland and South America) and the modern Australian lungfish *Neoceratodus forsteri*, a species that has survived virtually unchanged since the Cretaceous. Another species of *Neoceratodus*, *N. nargun*, is present in the Eumeralla Formation at Dinosaur Cove and Point Lewis on the southern Victorian coast. This form has delicate toothplates and ranges from the Cretaceous through to Pliocene.

Amphibians

Several jaws, a partial skull and a number of postcranial elements belonging to temnospondyl amphibians have been found in the Aptian Wonthaggi Formation of southern Victoria. These rare bones constitute the last definitive occurrence of temnospondyls in the fossil record and are a key example of relict-species survivorship in the cold high-latitude environments of south-eastern Australia during the Cretaceous. Temnospondyls reached a peak of diversity during the Triassic but had largely died out ~200 million years ago, leaving only isolated species from the Jurassic of Australia (see Chapter 3), central Asia and South Africa. The Australian Early Cretaceous temnospondyl has been named *Koolasuchus cleelandi*. It was a large-bodied (skull length ~650 mm, estimated body length up to 3 m) member of the Chigutisauridae. This group had a broad Gondwanan distribution during the Late Triassic and Early Jurassic; specimens are known from Australia, South America, India and possibly South Africa. *Koolasuchus* appears to have preyed on fish in inland rivers, a lifestyle that was adopted by crocodiles after the Aptian. The disappearance of the last temnospondyls has been linked to climatic warming across the Aptian-Albian boundary. Crocodiles, which are usually inactive below 10°C, may have outcompeted the cold-adapted temnospondyls once temperatures had increased sufficiently to allow them to survive.

Non-marine turtles

Non-marine turtle fossils are reasonably common in the Lower Cretaceous of Australia, with indeterminate cryptodires – turtles characterised by vertical retraction of the neck into the shell – reported from the Aptian Wonthaggi Formation of Victoria and various Albian units throughout eastern Australia. The smallest of the named species is *Chelycarapookus arctuatus*, described on the basis of an internal shell mould (~200 mm long) from the Albian Merino Group strata near Casterton in western Victoria. *Chelycarapookus* was a primitive turtle with a rounded shell similar to sinemydids, a group of Cretaceous cryptodirans from Asia. Unfortunately, the only known specimen of *Chelycarapookus* is too poorly preserved to allow a more conclusive examination and thus its relationships remain speculative.

A second cryptodiran, *Otwayemys cunicularius*, has been recovered from the lower Albian Eumeralla Formation at Dinosaur Cove in Victoria. This taxon is known from a shell, skull fragments and vertebrae representing several individuals. *Otwayemys* was about 20% larger than *Chelycarapookus* and may be related to meiolaniids, gigantic land turtles

A complete jaw of the relict Cretaceous temnospondyl amphibian *Koolasuchus cleelandi* from the Aptian Wonthaggi Formation of southern Victoria. Image courtesy Thomas Rich.

Opalised turtle and other reptile remains from the Lower Cretaceous Griman Creek Formation of Lightning Ridge. (Clockwise from top left) Skull of a terrestial meiolaniid and a cervical vertebra from a small aquatic turtle (images courtesy Robert Smith), the mandible fragment of *Crocodylus* (*Bottosaurus*) *selaslophensis*, and a piece of dermal bone from the skin of an unknown, possibly lizard-like animal.

with horns and tail clubs that lumbered across the Australian landscape and surrounding islands until only a few thousand years ago. Probable meiolaniid material has also been reported from the lower-middle Albian Griman Creek Formation at Lightning Ridge; it indicates an animal with a short broad snout, hoof-like claws and prominent triangular projections along the margin of the carapace. If confirmed, the identification of *Otwayemys* and the Lightning Ridge specimens as meiolaniid relatives implies an exceptionally long evolutionary history (middle Cretaceous to Holocene), and a unique radiation of primitive turtles on the Australian continent during the late Mesozoic.

Lizards

Australian Cretaceous lizard (lepidosaur) material is extremely rare. Reported specimens include an incomplete humerus from the Aptian Wonthaggi Formation near Cape Patterson in Victoria, thought to represent a large lizard of uncertain affinity, and some enigmatic sections of knobbly dermal bone from the Griman Creek Formation at Lightning Ridge, that have been attributed to an unknown lizard-like animal.

Crocodiles

Australian Lower Cretaceous crocodile material is fragmentary and restricted to deposits of Albian age. The Eumeralla Formation at Dinosaur Cove has yielded an isolated skull bone (quadratojugal) and several dermal scutes belonging to an unidentified crocodilian <2 m long. The Griman Creek Formation at Lightning Ridge has produced an incomplete mandible attributed to *Crocodylus* (*Bottosaurus*) *selaslophensis*. This specimen is unusual in having the teeth set into a common groove – a condition found nowhere else among crocodyliforms – making both its identification and the validity of its name doubtful.

Other indisputable crocodyliform bones and teeth have been collected from the opal-bearing clays at Lightning Ridge. These represent a broad-snouted form with laterally compressed teeth for shearing meat, and a second species with more conical dentition indicative of fish-eating habits.

Dinosaurs

Australian Early Cretaceous dinosaurs are known from footprints, a few partial skeletons and many isolated bones. Most of these finds are geographically and stratigraphically patchy but they have yielded sufficient information to indicate considerable faunal diversity and enigmatic biogeographical relationships. Berriasian dinosaur trackways are exposed on eroded wave-cut platforms along the coast near Broome in north-western Western Australia. Preserved in ancient wetland sediments of the Broome Sandstone (Carnarvon Basin), these abundant footprints represent a variety of dinosaurs

Dinosaur tracks in the Broome Sandstone exposed along the coast near Broome in north-western Western Australia. These deposits have yielded a variety of large dinosaur footprints including possible stegosaurs, a single footprint of which is shown here. Image courtesy John Long.

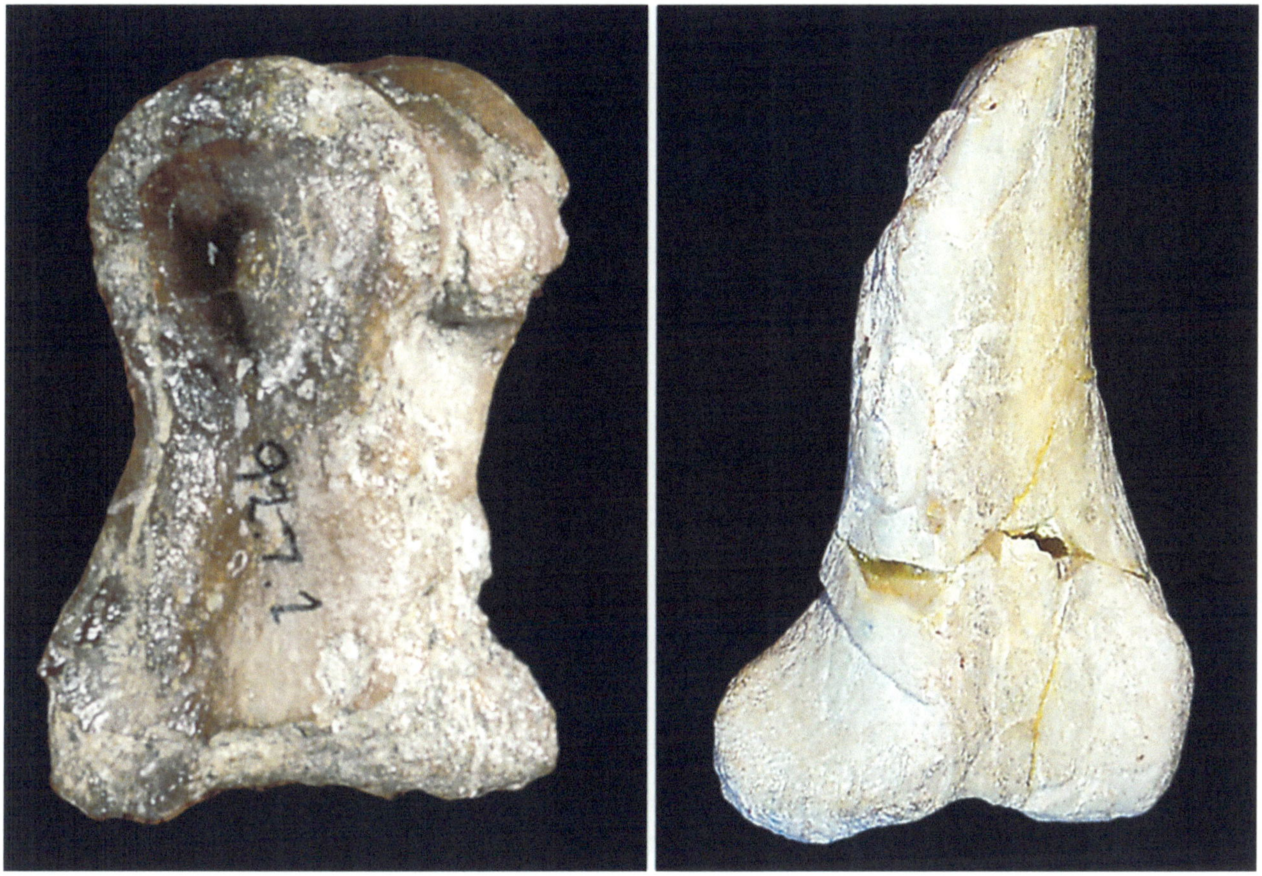

Lower Cretaceous theropod remains. (Left) An isolated phalanx from the Hauterivian-Barremian Birdrong Sandstone, Western Australia (image courtesy John Long). (Right) A distal tibia named *Kakuru kujani* from the Aptian Bulldog Shale, South Australia.

including sauropods (*Brontopus*, perhaps attributable to brachiosaurids), large (iguanodontian) and small (*Wintonopus*) ornithopods, large theropods (*Megalosauropus broomensis*) and possible stegosaurs (represented by a characteristic five-toed forefoot and three-toed hindfoot print). Evidence of Cretaceous stegosaurs is very rare, making the Broome Sandstone highly significant. Some of the herbivorous dinosaur tracks are also very large. Those of ornithopods are up to 800 mm in length and the prints of sauropods are occasionally 1.5 m across, suggesting animals over 40 m long!

The Hauterivian-Barremian marine rocks of the Birdrong Sandstone (Carnarvon Basin) near Kalbarri in Western Australia have produced a single theropod toe bone (phalanx). The Barremian-Aptian Wonthaggi Formation in the Gippsland Basin of southern Victoria has yielded a much more diverse array of taxa, including:

- numerous small herbivorous ornithopods *Qantassaurus intrepidus, Atlascopcosaurus loadsi, Fulgurotherium australae* and *Leaellynasaura amicagraphica* (the last three persisting into the Albian);
- an ankle bone (astragalus) of a primitive allosauroid, potentially referable to the recently described Australian theropod *Australovenator*;
- an potential ornithomimid (herbivorous ostrich-like dinosaurs);
- the ulna of *Serendipaceratops arthurcclarkei*, an apparent neoceratopsian (early horned dinosaurs like the famous *Protoceratops* from Mongolia).

Interestingly, ornithomimids and neoceratopsians are known elsewhere mainly from the Jurassic-Cretaceous of Asia and North America, thus implying periodic faunal exchanges between those

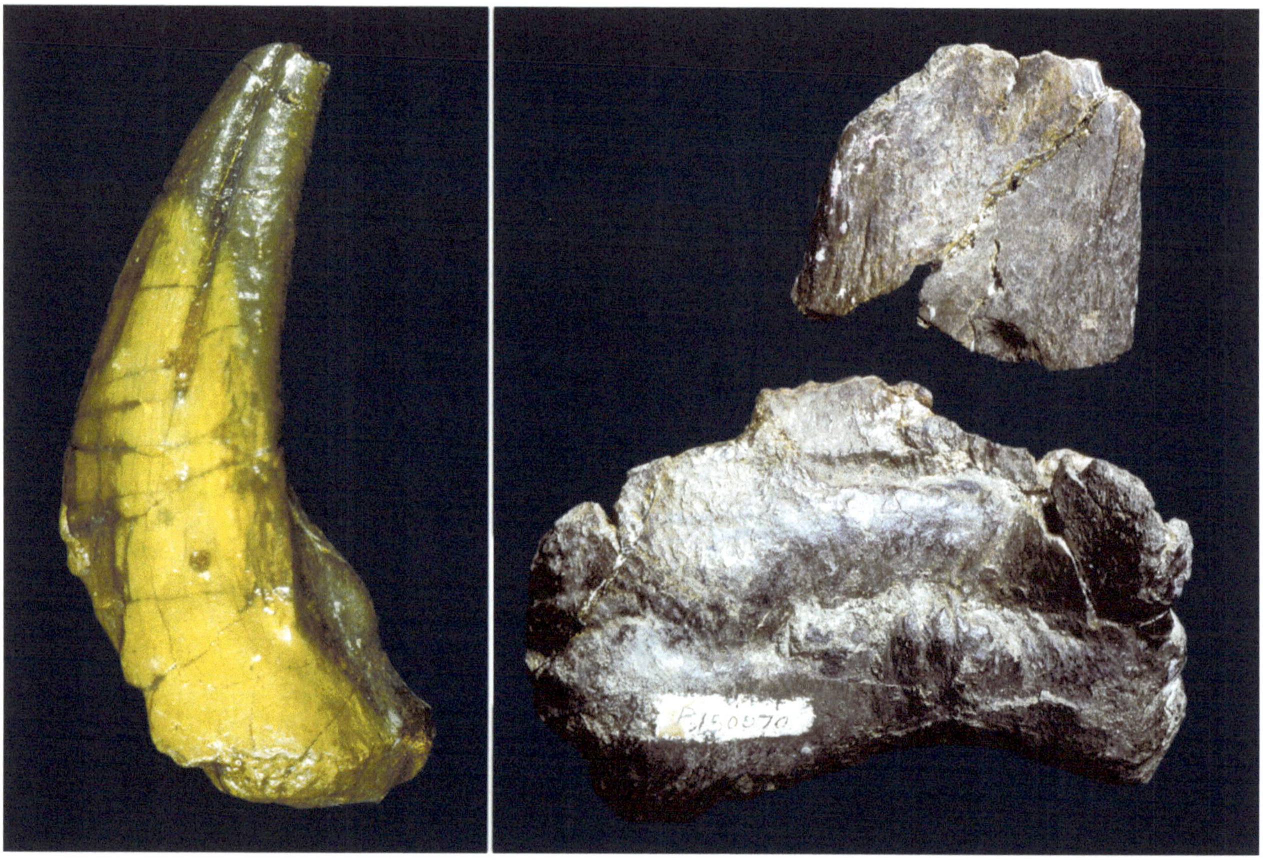

Aptian (Lower Cretaceous) theropod remains from southern Victoria. (Left) William Hamilton Ferguson's famous theropod claw, Australia's first dinosaur discovery. (Right) An allosauroid astragalus, possibly referable to *Australovenator.* Images courtesy Thomas Rich.

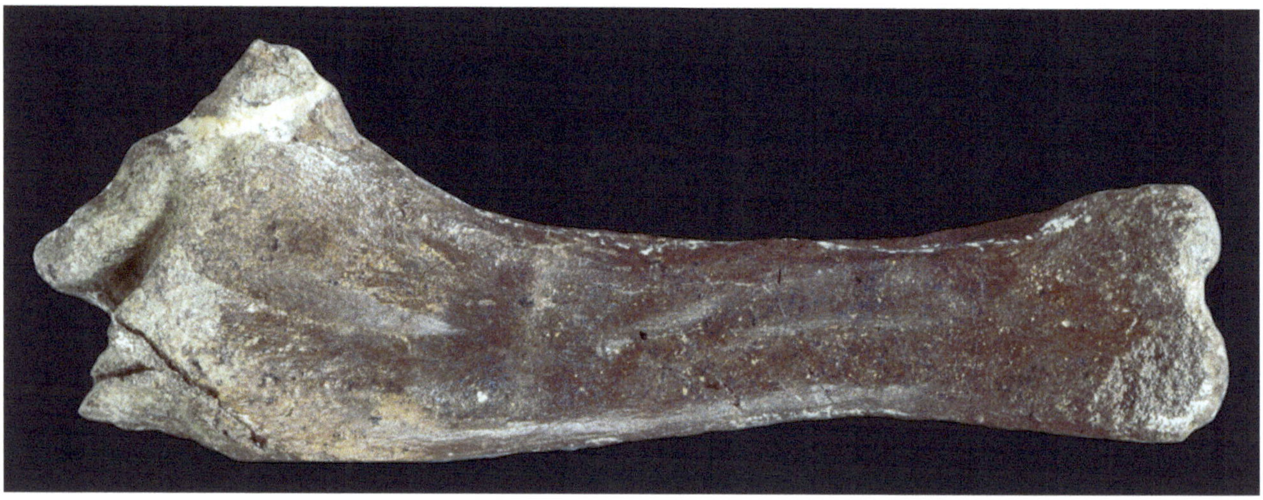

An isolated ulna attributed to the neoceratopsian *Serendipaceratops arthurcclarkei* from the Aptian Wonthaggi Formation of southern Victoria. Image courtesy Thomas Rich.

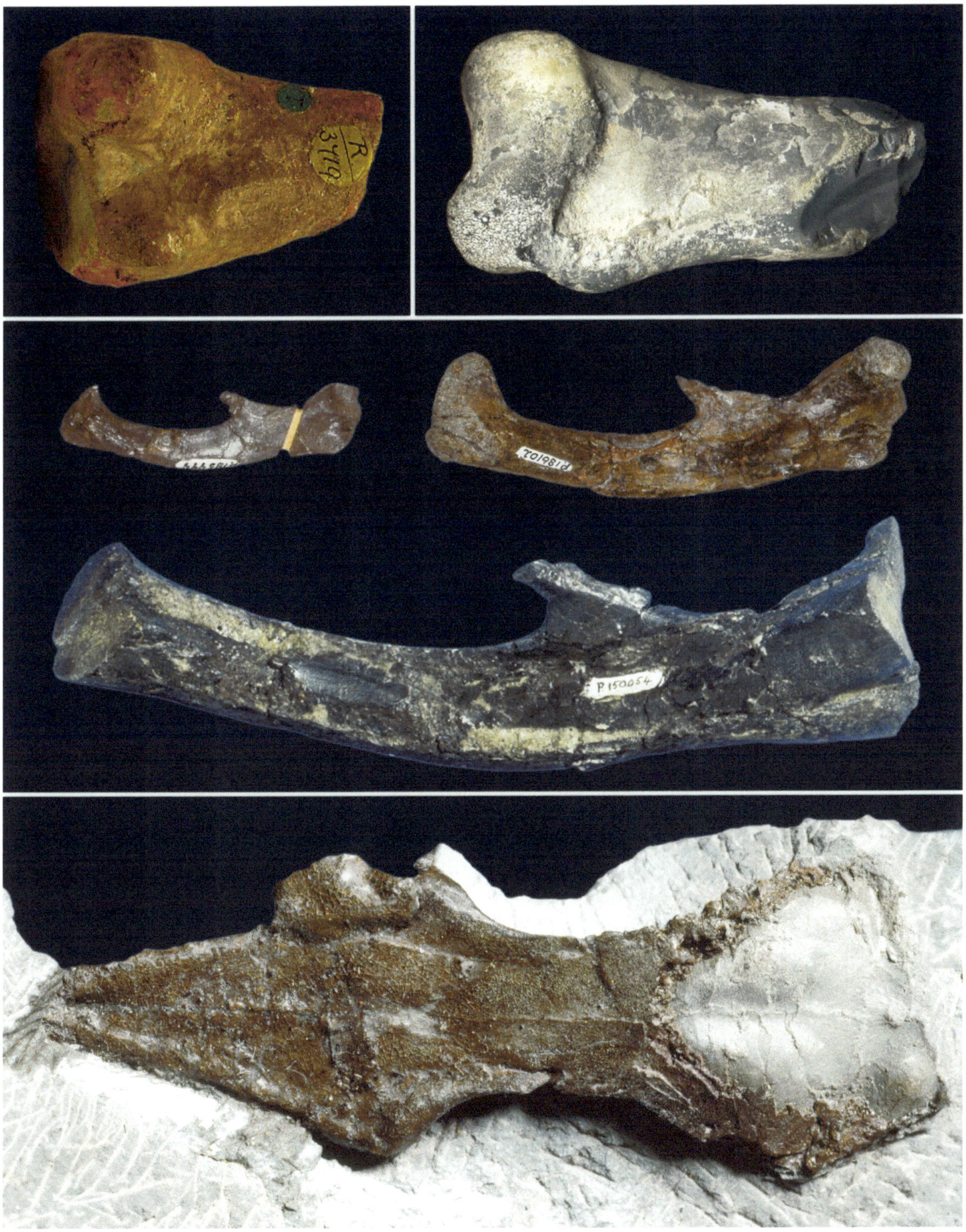

Small ornithopod remains from the Lower Cretaceous of Australia. (Clockwise from top left) The distal femur of *Fulgurotherium australae* and distal part of the tibia from another small ornithopod found at Lightning Ridge, several small ornithopod femora representing a growth series and the skull of *Leaellynasaura amicagraphica* from the Eumeralla Formation of southern Victoria. Images courtesy Thomas Rich.

landmasses and Australia during the Mesozoic. However, doubts have been raised about the affinities of some of these isolated bones. Alternative interpretations favour similarities to dinosaurs from Gondwana, particularly South America.

Other Aptian dinosaur fossils have been recovered from marine deposits of the Wallumbilla Formation at White Cliffs in New South Wales, the Bulldog Shale at Coober Pedy and Andamooka in South Australia, and the Bungil Formation near Roma in southern Queensland. The Wallumbilla Formation and Bulldog Shale remains comprise indeterminate medium-sized ornithopods (ankle and toe bones, Coober Pedy) and small theropods (limb bones, vertebra, phalanx, Andamooka; phalanx, White Cliffs). The only named taxon, *Kakuru kujani*, is a very bird-like theropod based on an opalised shinbone (tibia and fibula) and phalanx from Andamooka. The Bungil Formation has produced the holotype of *Minmi paravertebra*, an endemic primitive ankylosaur ~2–3 m in length with characteristic dermal armour of spines across the neck and a pavement of small circular ossicles embedded in the skin of its back and belly.

During the Albian, Australian dinosaur assemblages were dominated by small herbivorous ornithopods. *Atlascopcosaurus loadsi*, *Fulgurotherium australae*, *Leaellynasaura amicagraphica* and several unidentified species have been reported from the Eumeralla and Griman Creek formations. Large ornithopods include *Muttaburrasaurus*. This 7 m long primitive iguanodontian is known from the Griman Creek Formation (Lightning Ridge) and the marine Toolebuc Formation, Allaru Mudstone and Mackunda Formation of central Queensland. Only a single species, *Muttaburrasaurus langdoni*, has been described. It is distinguished by an inflated bony nasal structure on the snout that

Reconstruction of the small ornithopod *Leaellynasaura amicagraphica* from the Aptian-Albian deposits of Victoria and New South Wales. This chicken-sized dinosaur might have been specially adapted to the cool high-latitude climates of Early Cretaceous Australia.

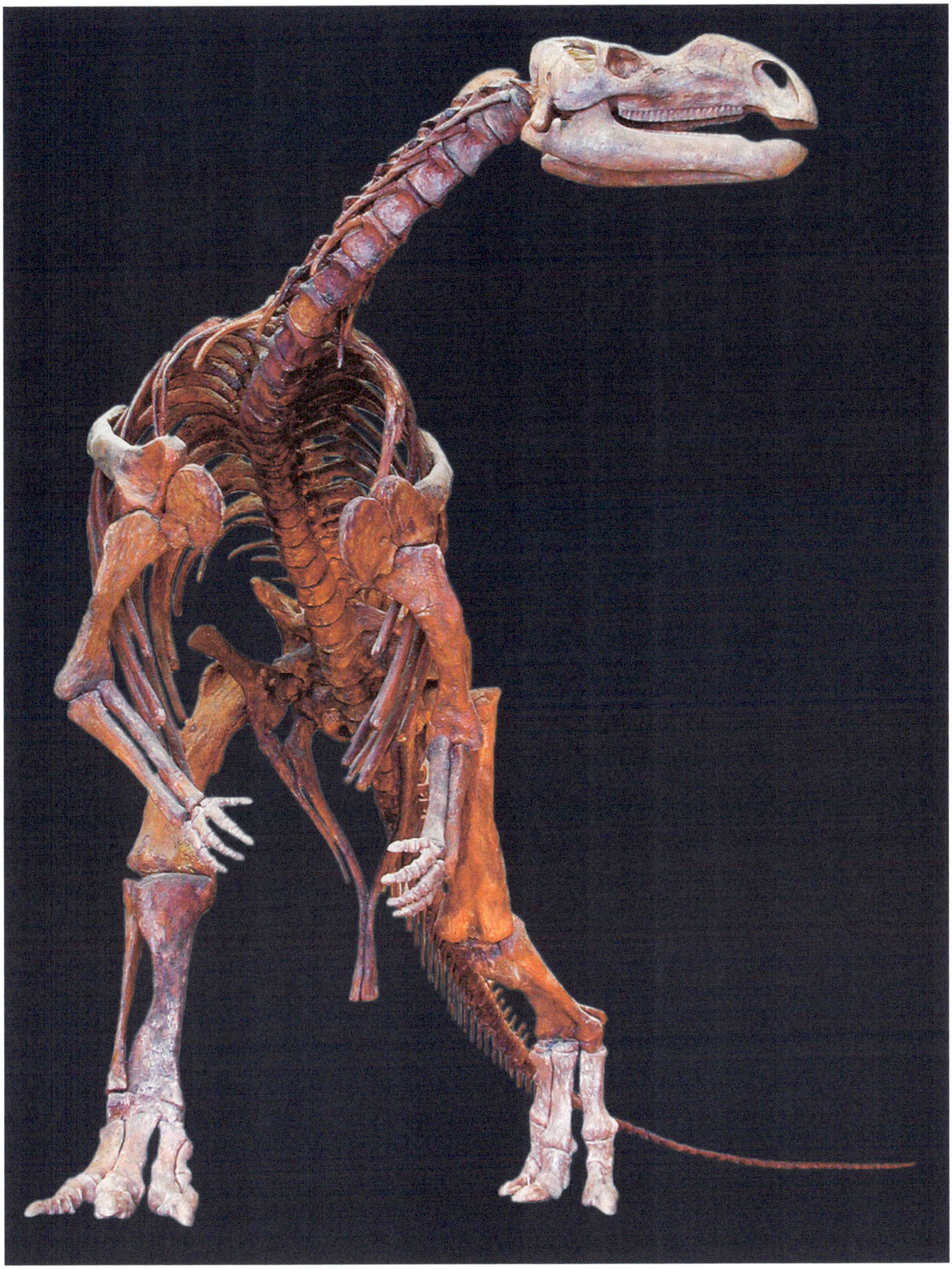

The complete reconstructed skeleton (as mounted in the Queensland Museum, Brisbane) of one of Australia's best-known dinosaurs, *Muttaburrasaurus langdoni*. Remains of this bus-sized herbivore have been recovered from a number of Albian rock units throughout eastern Australia.

(Top) Reconstruction of the sheep-sized ankylosaur *Minmi paravertebra.* (Bottom) The distinctive armoured skin comprises bony ossicles and large scutes that are occasionally preserved in situ. This fragment is part of a near-complete skeleton from the Albian sediments near Richmond in central Queensland.

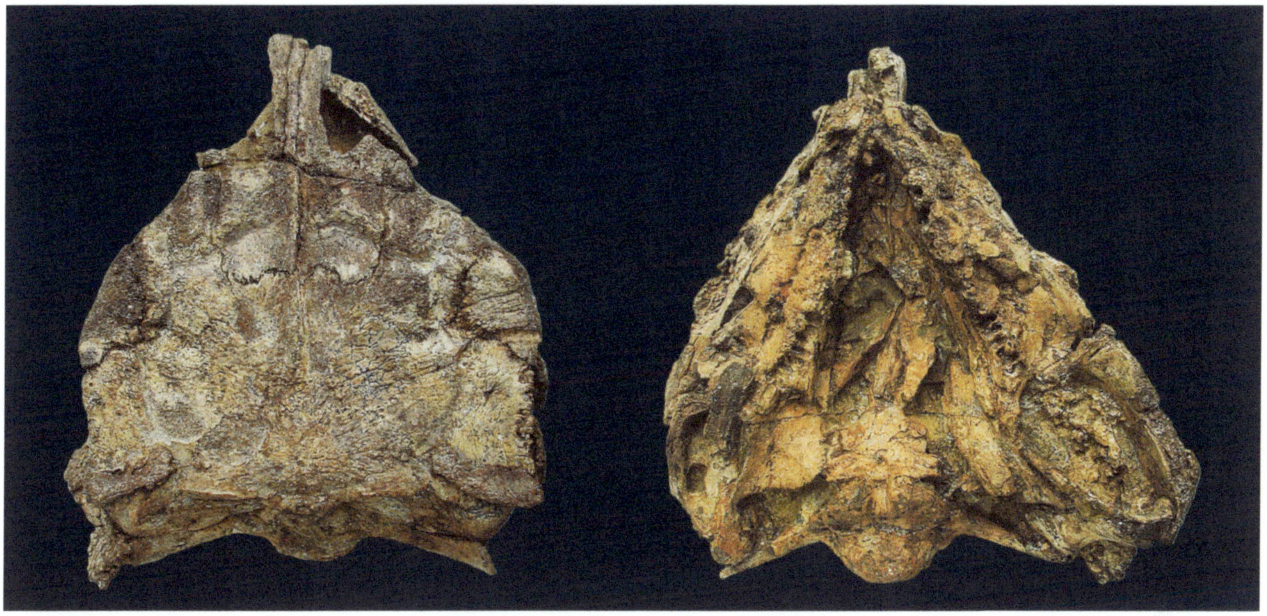

Skull of the primitive Aptian-Albian ankylosaur *Minmi* shown in (left) dorsal and (right) palatal views. This specimen is from the Albian marine deposits around Richmond in central Queensland.

might have been used to amplify its calls and perhaps varied in size between male and female individuals. *Muttaburrasaurus* was primarily quadrupedal, but probably reared up on its hindlegs to browse taller vegetation. Its blade-like teeth formed a shearing edge, which differs from more advanced iguanodontians whose teeth were arranged into a broad grinding surface.

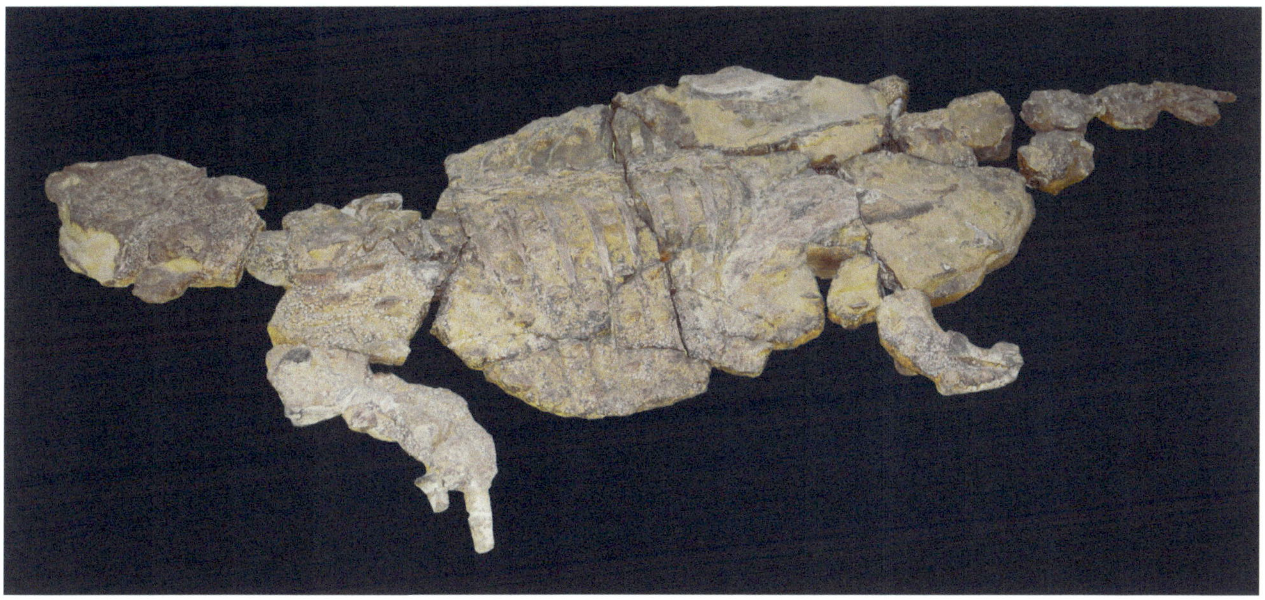

The partial skeleton of *Minmi* found in Albian marine deposits near Richmond in central Queensland. This is probably the most complete single dinosaur specimen ever found in Australia.

A group of dromaeosaur-like theropods, including several chicks, scavenge turtle and fish carcasses in a dried riverbed near what is now Lightning Ridge in northern New South Wales. The deposits at Lightning Ridge have been a productive source of terrestrial plant and animal fossils dating from ~110 million years ago.

The Toolebuc Formation and Allaru Mudstone have also produced specimens of the ankylosaur *Minmi*, including one superbly preserved skeleton with the skull and remains of its last meal – chewed-up fragments of twigs and leaves. Isolated bones and armour scutes of ankylosaurs have been found in the Eumeralla and Griman Creek formations but are too incomplete to identify more specifically. The Eumeralla Formation has yielded a second possible neoceratopsian ulna, which is smaller and more gracile than that of *Serendipaceratops*.

Albian sauropods are known from the Toolebuc Formation, Allaru Mudstone and Griman Creek Formation. The Allaru Mudstone material has been referred to *Austrosaurus mckillopi*, an ambiguous titanosaurian based on a few incomplete dorsal vertebrae. Isolated teeth and tail vertebrae from the Griman Creek Formation at Lightning Ridge are also indicative of titanosauriforms. A partial neck vertebra from the Toolebuc Formation near Hughenden in central Queensland might be a brachiosaurid.

Australian Albian theropod remains include isolated bones of a possible oviraptorosaur (caenagnathid), a single ulna resembling the South American theropod *Megaraptor*, and a thighbone (femur) of the putative ornithomimid *Timimus hermani*, all from the Eumeralla Formation around Dinosaur Cove. Like a number of other Victorian dinosaurs, these are controversial – some researchers prefer indeterminate assigment to higher-level groupings. Numerous teeth and a large claw attributable to dromaeosaurid-like dinosaurs (similar to the famous *Velociraptor*) have been recovered from the Eumeralla Formation and Griman Creek Formation. The Griman Creek Formation has produced an opalised theropod metatarsal ~250 mm long, the tail vertebra of *Walgettosuchus woodwardi* (now considered taxonomically invalid) and an unusual finger bone (metacarpal) dubbed *Rapator ornitholestoides*, which represents a predator ~5 m long. It was named because of its structural similarity to the primitive North American coelurosaurian *Ornitholestes*.

Pterosaurs

Pterosaurs (Pterosauria, often referred to as pterodactyls) were flying reptiles, widespread from the Late Triassic until the end of the Cretaceous. Their wings comprised a membrane of skin stretched from the hindleg to the tip of the elongated fourth finger. The earliest pterosaurs (rhamphorhynchoids) were generally small, with toothed jaws and long tails that bore membranous rudder at the tip. More advanced forms (pterodactyloids) are typified by a stumpy tail, reduction or loss of the teeth and occasionally gigantic size.

Australian pterosaur fossils are extremely rare, but a small collection of material has been recovered from the Lower Cretaceous marine Toolebuc Formation (Albian) near the townships of Boulia and Hughenden in western Queensland. The specimens include jaw fragments, vertebrae, wing bones, a shoulder girdle (scapulocoracoid) and pelvis attributable to ornithocheiroids. This was a group of pterodactyloids that incorporated the famous crested *Pteranodon* and had a broad distribution encompassing South America, Asia, Europe and North America.

The only named Australian pterosaur was based on a partial skull with very robust, widely spaced, conical teeth. Dubbed *Mythunga camara*, the relationships of this animal are equivocal but appear to lie with ornithocheiroids. *Mythunga* was probably a fish-eater and it may have had a wingspan of 4 m.

Pterosaur bones have been found in the Albian Eumeralla Formation and the Griman Creek Formation of Victoria and New South Wales respectively. The remains are non-diagnostic, but they indicate that pterodactyloids probably lived in coastal estuaries and inland regions bordering the Australian epicontinental seaway.

Birds

Australian Mesozoic bird (avian) fossils are fragmentary and known only from Lower Cretaceous deposits. The earliest recorded finds include several small contour (body) feather impressions from Aptian lacustrine clays of the Korumburra Group near Koonwarra in southern Victoria. Avian body fossils occur in the Aptian Wonthaggi Formation deposits near Inverloch (Flat Rocks Site) in Victoria (an isolated furcula, or wishbone) and the Albian Griman Creek Formation of Lightning Ridge, New South Wales. The Lightning Ridge material indicates both a small-bodied gull-like form with

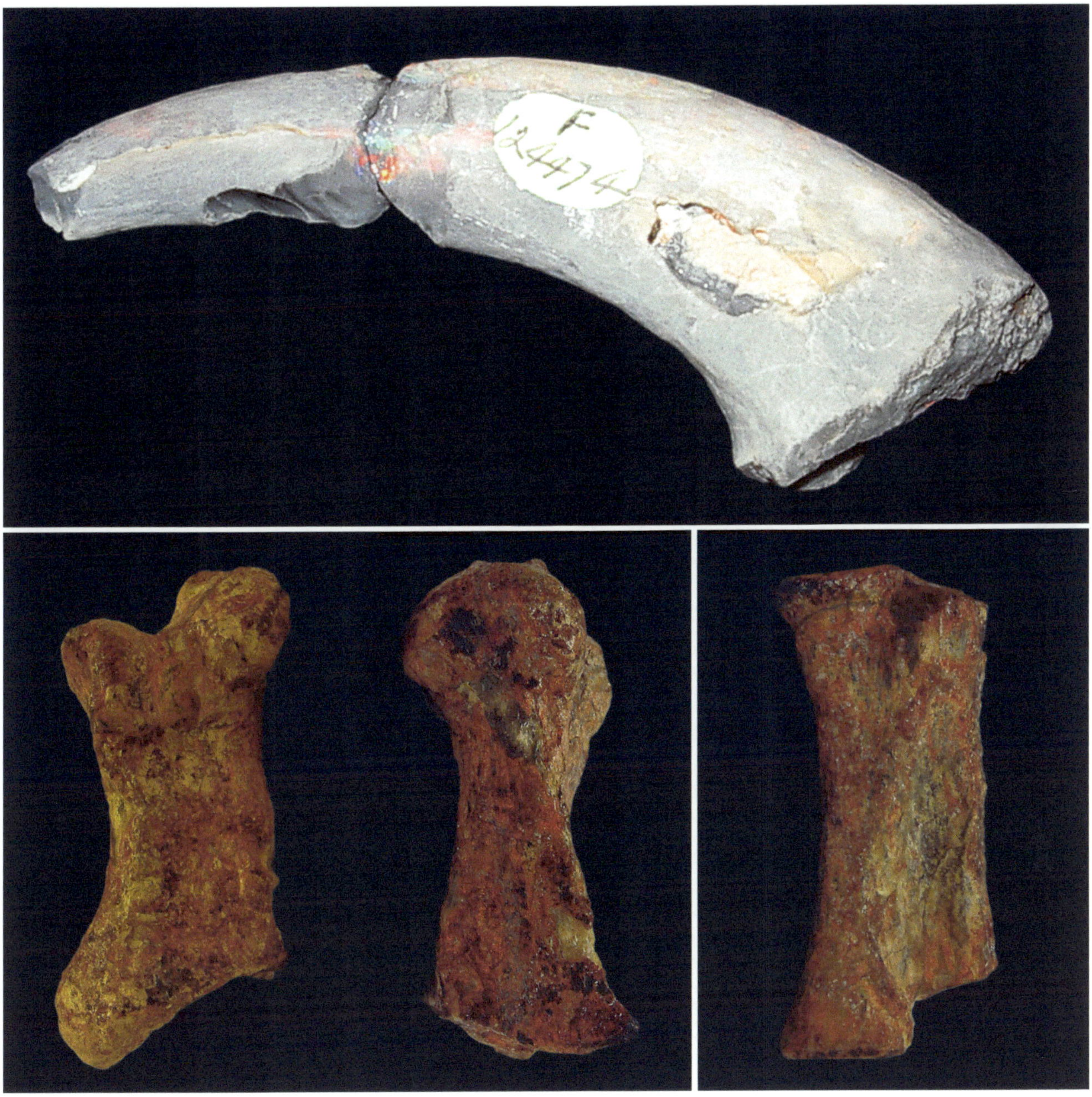

Albian (Lower Cretaceous) theropod remains from the Griman Creek Formation at Lightning Ridge, New South Wales. (Clockwise from top) A dromaeosaur-like toe claw (image courtesy Henk Godthelp), a caudal centrum attributed to the now-invalid taxon *Wallgettosuchus woodwardi*, and the metacarpal of *Rapator ornitholestoides*.

vertebral morphology comparable to the toothed ornithomorph *Ichthyornis* (well-known from western Asia and North America), and a larger diving bird represented by limb bones (tibiotarsi) with shaft dimensions similar to the flightless marine *Baptornis* and *Hesperornis* from the Late Cretaceous of North America.

Other Australian Mesozoic bird fossils have been found in the upper Albian Toolebuc Formation around Boulia in western Queensland, including limb bones and vertebral elements attributed to the sparrow-sized enantiornithine *Nanantius eos*. Enantiornithines are a primitive avian group with widespread Gondwanan distribution. Most enantiornithines are

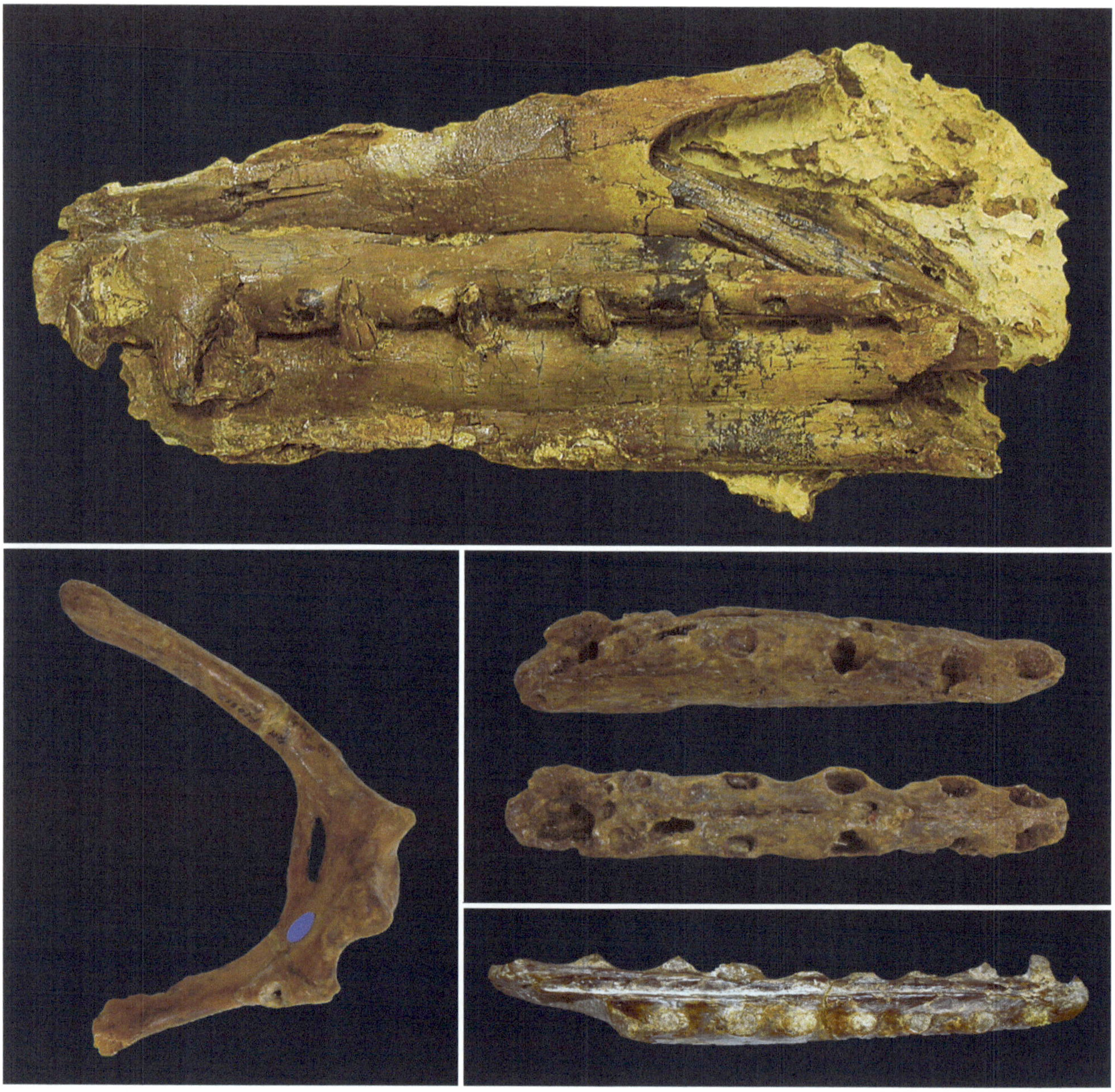

Pterosaur remains from the Albian Toolebuc Formation of central Queensland. (Clockwise from top) The partial skull of the only named Australian pterosaur, *Mythunga camara*, found near Hughenden, two views of the tip of a lower jaw from Boulia (thought to be from an ornithocheiroid), the lower jaw of an ornithocheiroid found near Hughenden, and a scapulocoracoid from Boulia.

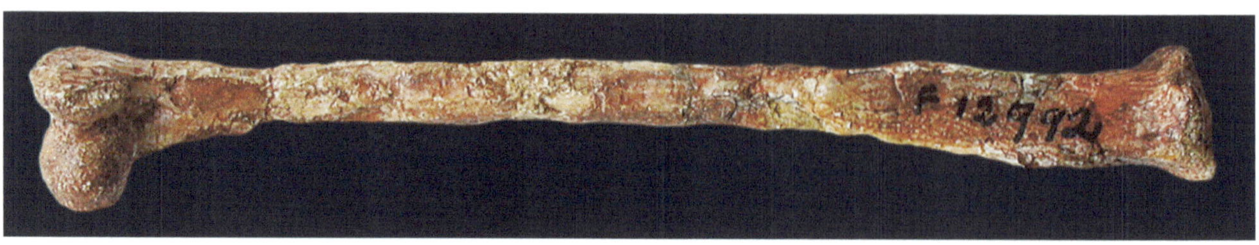

Tibiotarsus of the primitive enantiornithine bird *Nanantius eos* from the Albian Toolebuc Formation of Queensland.

thought to have been woodland inhabitants; the Toolebuc Formation remains were probably transported into marine environments as carcasses washed out to sea from nearby coastal habitats.

Relict therapsids

The unexpected discovery of what appear to be dicynodont skull fragments was recently reported from the upper Albian marine Allaru Mudstone near the township of Hughenden in northern-central Queensland. The hippopotamus-like herbivores lacked teeth, but probably used their horny beak and tusks to dig for roots and tubers. The last known dicynodonts come from the Upper Triassic (Norian, ~215 million years ago), thus predating the Lower Cretaceous material by over 100 million years. Nevertheless, the Australian specimen displays a number of features diagnostic for the group including open-rooted enamel-less tusks with dentine layers stacked in a cone-in-cone structure; this feature is unique to dicynodonts. The presence of a relict dicynodont, together with temnospondyl amphibians and other anachronistic vertebrates in the Lower Cretaceous deposits of Australia, provides further evidence that the continent's terrestrial faunas were isolated, perhaps via climatic barriers, throughout much of the Mesozoic.

Cretaceous mammals

The record of Lower Cretaceous mammals in Australia is limited to deposits of Aptian-Albian age. All of the Aptian material is derived from the Wonthaggi Formation at Flat Rocks near Inverloch in Victoria. Four taxa have been identified: *Corriebaatar marywaltersae*, *Ausktribosphenos nyktos*, *Bishops whitmorei* and *Teinolophos trusleri*. These mouse-sized (skull lengths of 25–30 mm) mammals were established on the basis of isolated jaws and they represent separate lineages. The most recent and perhaps most unexpected find, *Corriebaatar marywaltersae*, is a multituberculate – a cosmopolitan group of mainly Mesozoic mammals whose fossils are well-known from Cretaceous deposits in the northern hemisphere. The recognition of *Corriebaatar*, which also constitutes its own unique family Corriebataaridae, is highly significant because it predates other Late Cretaceous Gondwanan multituberculates from South America and thus could indicate a migration of these mammals into Australia prior climatic cooling at the southern pole.

Ausktribosphenos and *Bishops* have been controversially described as placentals (or eutherians). The earliest known placentals come from Asia (Barremian, ~125–130 million years old); identification of these two primitive Australian forms implies that the group was present in the region over 60 million years earlier than previously suspected. Before this, the oldest definitive Australian placentals were bats from the Palaeogene of Murgon in Queensland, ~54 million years old. *Ausktribosphenos* and *Bishops* were assigned to Eutheria on the basis of characteristic wear patterns on the teeth, the presence of three molars with a molariform last premolar (marsupials possess four molars and a non-molariform premolar), and the structure of the jaw joint. However, other work has suggested that they are distinct enough to warrant their own radiation – the

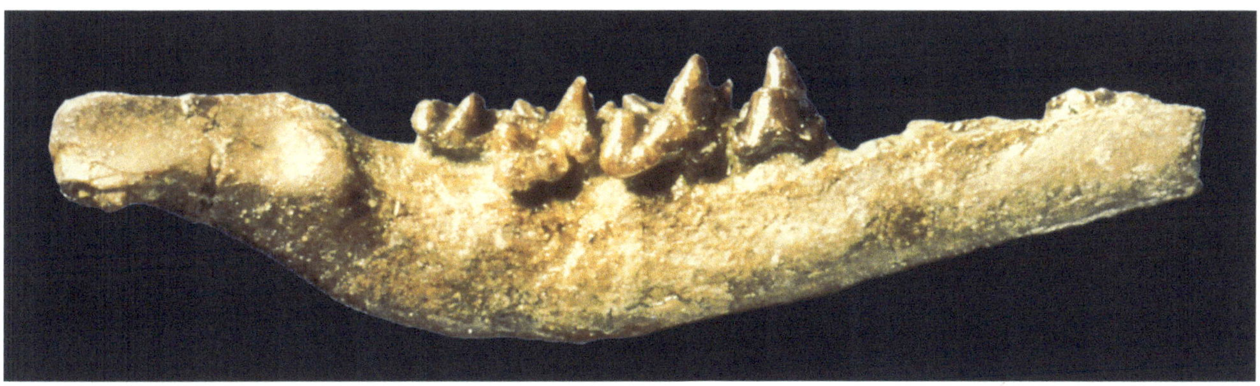

Tiny jaw of the possible placental mammal *Ausktribosphenos nyktos* from the Aptian Wonthaggi Formation of southern Victoria. Image courtesy Thomas Rich.

Ausktribosphenida – and should be placed with monotremes in the Australosphenida, an exclusively Gondwanan clade distinct from placentals and marsupials (which constitute a northern hemisphere radiation, the Boreosphenida).

The fourth Wonthaggi Formation mammal, *Teinolophos trusleri*, is a member of the Monotremata. This group incorporates the Albian taxa *Kryoryctes cadburyi*, from the Eumeralla Formation, and *Steropodon galmani* and *Kollikodon ritchiei* from the Griman Creek Formation. Monotremes probably evolved in the Australian region during the Mesozoic and spread to South America via Antarctica in the latest Cretaceous (platypus-like ornithorhynchid monotremes are known from the earliest Palaeogene of Patagonia). *Teinolophos trusleri* is the smallest

Australian Lower Cretaceous monotreme fossils. (Clockwise from top) Upper jaw of the bizarre monotreme *Kollikodon ritchiei* (image courtesy Andrew Cody), jaw of *Steropodon galmani* and the humerus of *Kryoryctes cadburyi* (image courtesy Thomas Rich).

monotreme on record (mandible length 25 mm), and has an unusual jaw that suggests a powerful bite. Conversely, Albian monotreme taxa are large-bodied (~400 mm long) and appear to have been terrestrial burrowers or aquatic. *Kryoryctes* is based on an incomplete humerus that is structurally similar to those of modern echidnas (Tachyglossidae), but it appears to be a more primitive monotreme with a comparable burrowing lifestyle. *Steropodon* and *Kollikodon*, both represented by jaws from the opal mines of Lightning Ridge, are thought to be aquatic. *Steropodon* has been placed with *Teinolophos* as the most primitive monotremes and has a molar structure similar to ornithorhynchids. *Kollikodon* forms a separate grouping (Kollikodontidae) based on its unusual molars, which were developed into bulbous crushing plates. *Kollikodon* probably fed on hard-shelled invertebrates (molluscs) and may have detected its prey using electrosensory organs on the snout.

Recently, the Griman Creek Formation and near-shore marine Toolebuc Formation of north-western Queensland have produced more unusual mammal fossils. The Griman Creek Formation remains include a bizarre tooth from what may be a dryolestid and a large incomplete jaw (probably ~80 mm long when complete) with sockets for three molars, similar to that of ausktribosphenids. The Toolebuc Formation material consists of a tiny marsupial-like vertebra and an incisor tooth. None can be diagnosed further at this stage but their recognition indicates that much Australian Mesozoic mammal material still awaits discovery.

7

AFTER THE OCEAN: LIFE FROM THE LATE CRETACEOUS

Around 100 million years ago (latest Albian to Cenomanian) the great inland sea started to drain off the Australian continent. Left in its wake were vast new areas of lowland lakes and swamps covering much of central and north-eastern Australia. Areas of marine inundation persisted along the western and south-western coastal margins (modern Victoria) but these receded by the Turonian (93 million years ago), leaving only shallow embayments along the Western Australian coast. In the north, the Darwin region and nearby offshore islands (Bathurst and Melville islands) remained underwater until the Turonian, depositing sediments of the Bathurst Island Group. After the Turonian, they and the smaller landmasses further north (Timor, Moluccas, Celebes) became exposed as dry land. This widespread seaway regression is generally attributed to active crustal movements, in particular the commencement of circum-global mountain-building events that continued into the Cainozoic.

Australian Upper Cretaceous fossils are best-known from Western Australia and central Queensland, and preserve evidence of both marine and terrestrial assemblages. In Western Australia, the remains represent shallow marine faunas that occupied the western margins of the emerging Indian Ocean. Most of the fossils are derived from limestone, marl and chalk (a friable rock composed of shells from unicellular plankton called foraminifera) exposed along the coast. Some of the most notable deposits are the Molecap Greensand, Gingin Chalk and Miria Formation, which crop out respectively in the Gingin district (Perth Basin) north of Perth, the north-western Murchison River region and the Cardabia-Giralia Ranges in the eastern and central parts of the Carnarvon Basin (north-western Western Australia). The depth of these sequences is over 100 m around Gingin, and several hundred metres in the Carnarvon Basin. In many places the rocks are extremely fossiliferous and provide a highly detailed record of marine life in the Australian region close to the very end of the Age of Dinosaurs (Maastrichtian, ~65 million years ago).

In contrast, knowledge of Australia's Upper Cretaceous terrestrial faunas and floras is largely restricted to a single rock unit, the Winton Formation. This comprises extensive siltstone/sandstone deposits across the Eromanga Basin through central-western Queensland and into South Australia. They represent sediments laid down by the vast network of rivers and streams that covered the lowland plains exposed after the retreat of the inland sea ~100 million years ago. Key localities, such as those around Winton in central Queensland, have produced a wealth of fossils including dinosaurs and early flowering plants. The remainder of the Australian Upper Cretaceous terrestrial record is extremely poor (a product of limited rock outcrop), with a few occurrences of terrestrial vertebrate remains and microfossils (spores and pollen) reported from marine rocks in Western Australia.

Excavations in the Albian-Cenomanian Winton Formation deposits in central and south-western Queensland have yielded a diversity of non-marine tetrapods such as dinosaurs, turtles, crocodiles and lungfish. Plant remains are also extremely common. Image courtesy Michael Lee.

Significant Australian Upper Cretaceous deposits and localities

The Bathurst Island Group contains the oldest of Australia's Upper Cretaceous marine sequences, and spans the transition between the Early Cretaceous (Aptian-Albian) inland sea and its final regression from the Australian continent during the mid-Late Cretaceous (Cenomanian-Turonian). The Bathurst Island Group comprises a continuous series of mudstone and glauconitic siltstone/sandstone units that are widely distributed across the northern Australian Money Shoals Platform, particularly in the coastal areas around Darwin (e.g. Casuarina Beach), the Cobourg Peninsula and Bathurst and Melville islands. The Darwin region deposits are mainly Lower Cretaceous in age. They include the Darwin Formation and the laterally equivalent Marligar Formation. The Upper Cretaceous strata are best exposed around the Mountnorris Bay area (Cobourg Peninsula) and Bathurst and Melville islands. They include the conformably overlying Walangarlu Mudstone and Moonkinu Formation, units that record a regressive event along the then-emerging northern coastal margin of Australia – low-energy deeper-water environments (Walangarlu Mudstone) were steadily replaced by higher-energy shallower-water conditions (Moonkinu Formation). The age of the Walangarlu Mudstone/Moonkinu Formation has been determined on the basis of mollusc assemblages principally comprising ammonites, plus bivalves, gastropods and scaphopods. These indicate an uppermost Albian to Cenomanian age for the Walangarlu Mudstone (possibly extending into the Lower Albian at the very base of the unit), and Cenomanian-Turonian age for the overlying Moonkinu Formation.

Slightly younger Upper Cretaceous marine rocks are found along the Western Australian coast. The Molecap Greensand and overlying Gingin Chalk comprise a continuous sequence of weathered

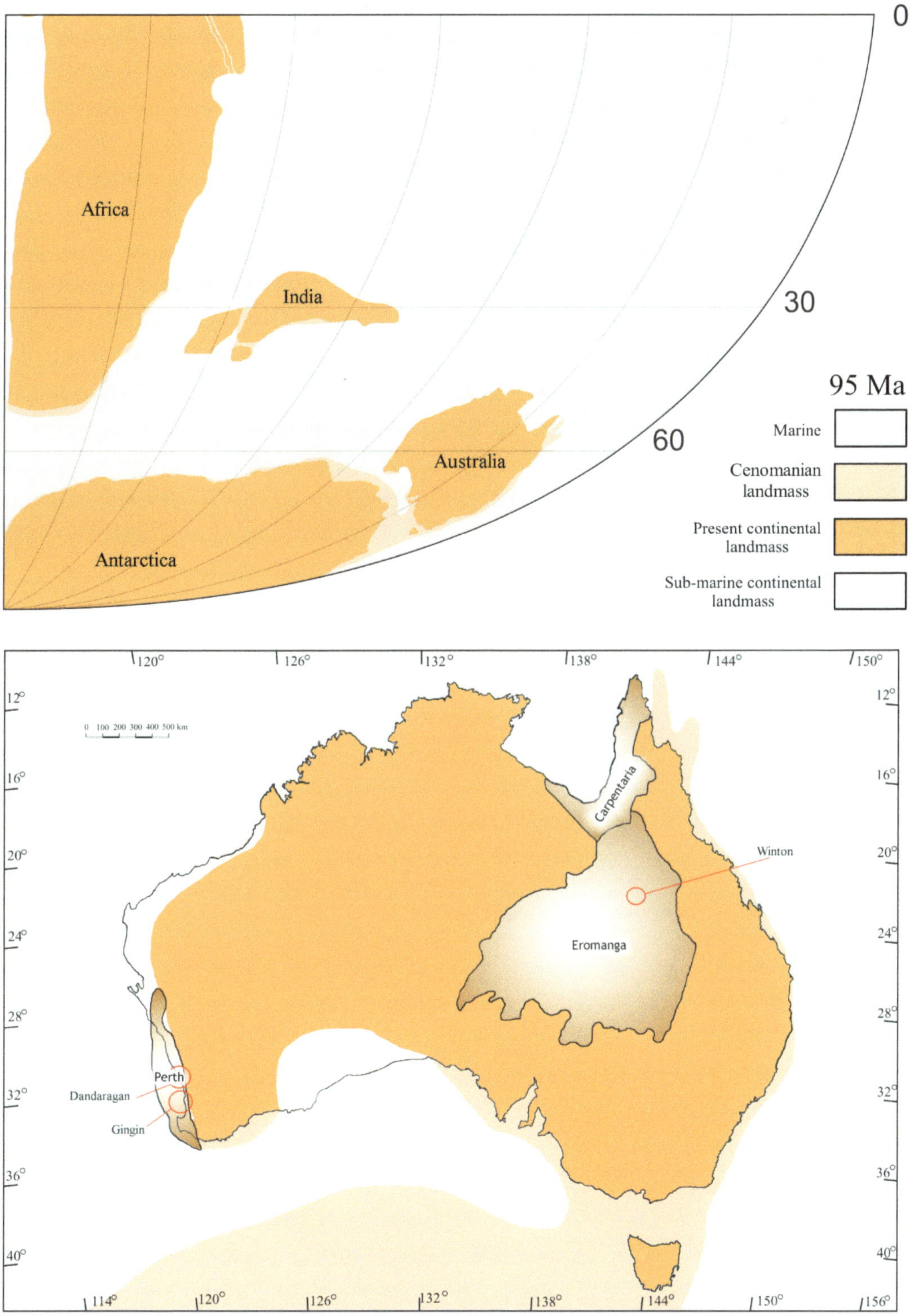

(Top) The geographical position of the Australian continent from the Cenomanian stage of the Late Cretaceous, ~95 million years ago. (Bottom) Principal marine fossil localities from this time-frame (indicated in red) occur around the townships of Dandaragan and Gingin in the Perth Basin of Western Australia. Terrestrial fossils of a similar age have been recovered from the Winton region of central Queensland.

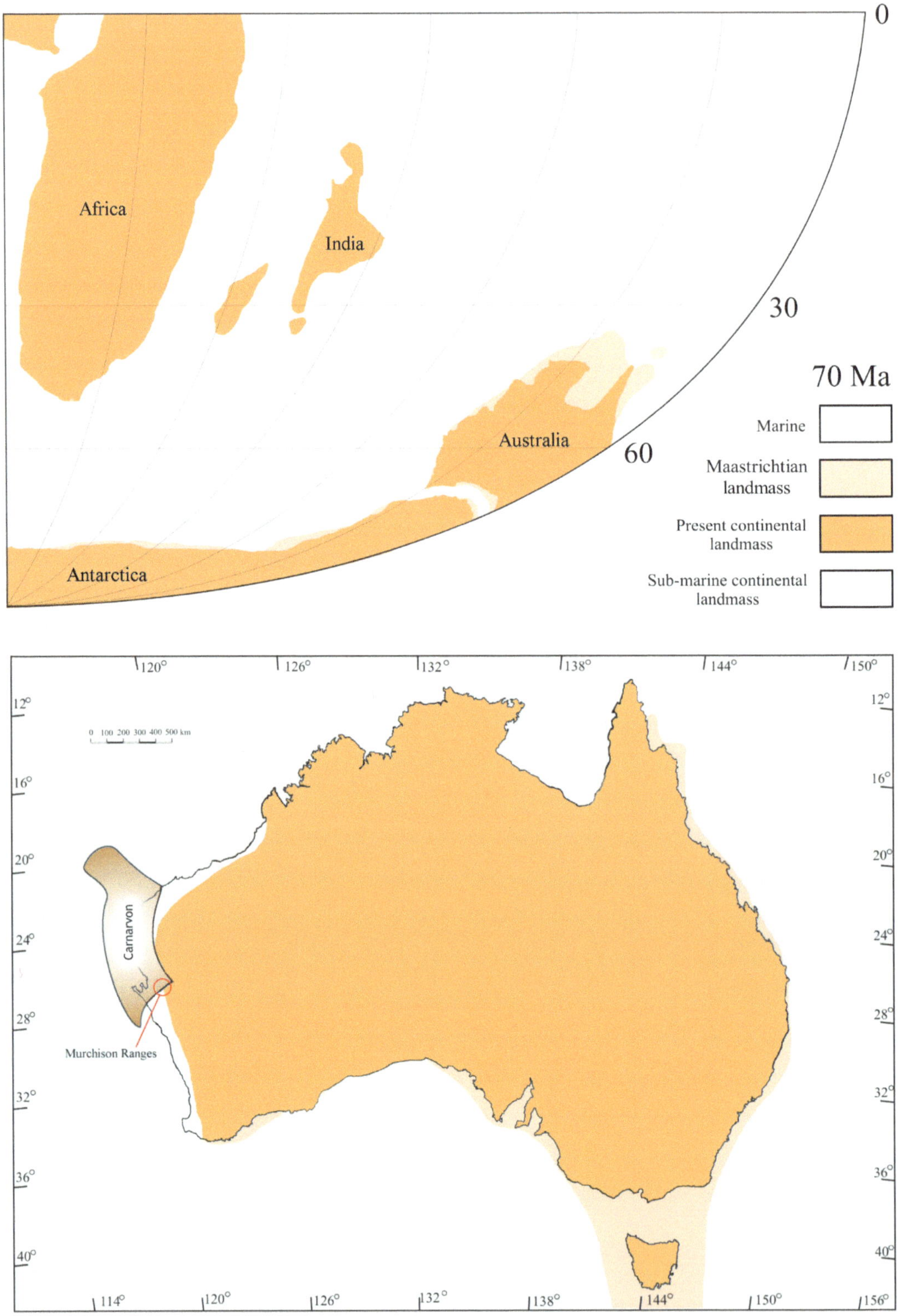

(Top) The geographical position of the Australian continent during the Maastrichtian stage of the Late Cretaceous, ~70 million years ago. (Bottom) Maastrichtian fossil localities occur in the Murchison Ranges (indicated in red) of the Carnarvon Basin of Western Australia.

shallow marine strata that outcrop extensively around the townships of Dandaragan and Gingin (Perth Basin) in south-western Western Australia. The Molecap Greensand comprises mainly glauconitic sandstones that lack distinct bedding and may contain considerable amounts of reworked material. In contrast, the overlying Gingin Chalk is composed of extremely fine chalky sediments formed by the compaction of planktonic skeletons. The palaeoenvironmental setting for both units is thought to have been shallow coastal marine, with deposition under poorly oxygenated bottom-water conditions.

The age of the Molecap Greensand is contentious. Rich palynomorph assemblages indicate an unusual mixed assemblage of Cenomanian-Coniacian plankton, possibly the result of sediment slumping associated with the Yallalie impact structure – a buried Late Cretaceous asteroid crater. The Gingin Chalk is variously considered Santonian through to lowermost Campanian.

The Molecap Greensand/Gingin Chalk macrofossil record is fragmentary but includes evidence of marine reptiles such as mosasaurs, plesiosaurs and ichthyosaurs, as well as sharks, bony fish and dinosaurs. Invertebrate material, dominated by bottom-living molluscs, is common in the Gingin Chalk.

The stratigraphically youngest fossil-bearing Cretaceous rocks in Australia include the Miria Formation. This unit comprises thin sequences of uncemented chalk (calcarenite) sediments, laid down on a calm continental shelf created during a Late Cretaceous marine incursion into the Carnarvon Basin of north-western Western Australia. Macrofossils are plentiful in the Miria Formation but specimens are frequently incomplete. Storm activity may have broken them up at the time of deposition.

The Miria Formation is typically poorly exposed but its outcrops can be recognised by phosphatic steinkerns that occur within weathered sediments. These are most conspicuous in the Giralia Ranges south of Exmouth Gulf, and they often contain fossils. Faunal records for the Miria Formation indicate an extremely diverse range of pelagic cephalopods and benthic molluscs. Mosasaurs, sharks and isolated dinosaur and pterosaur bones have also been found. Ammonites and microfossils suggest a Late Maastrichtian age for the unit. Plankton assemblages denote warm temperate conditions.

The Australian Upper Cretaceous terrestrial fossil record is derived almost entirely from the Winton Formation. This is composed of claystones, mudstones, coal and volcanically derived sandstones that extend over a broad area of the Eromanga Basin in south-western Queensland and north-eastern South Australia. At its maximum, the Winton Formation is an amazing 1.2 km thick, but low topographic relief, deep weathering and oxidised ironstone cover often make identification of surface outcrops difficult. The Winton Formation forms the uppermost section of the Rolling Downs Group and is considered uppermost Albian to lower Cenomanian in age, on the basis of invertebrate and microfossil assemblages. However, fossil pollen records from the uppermost beds imply that deposition may have continued into the late Cenomanian or early Turonian, at least in the central part of the basin (central Queensland).

The Winton Formation is interpreted as a floodplain deposit generated by river systems draining inland towards permanent lakes and swamps. Macrofossils include freshwater unionoid bivalves, insects, aquatic vertebrates (fish and reptiles) and dinosaurs (known from body fossils and trackways).

Land plants

Australian Late Cretaceous terrestrial vegetation is known from impressions of leaves, stems and reproductive structures (cones), mineralised wood, cuticles, spores and pollen. Much of the record is geographically and stratigraphically patchy, but is nonetheless evidence of major changes in rainforest and woodland communities.

Uppermost Albian-Cenomanian macrofloras are derived from the Winton Formation and include forest and open woodland canopy species. The dominant vegetation types are mainly conifers, including several araucarians (*Araucaria*), a podocarp (*Elatocladus*), *Ginkgo* and sequoias (Taxodiaceae) such as *Athrotaxus* and *Australosequoia wintonensis*. Probable understorey plants are pentoxylaceans (*Taeniopteris*) and ferns (e.g. *Sphenopteris* and *Microphyllopteris*), with horsetails and quillworts at swamp and lake margins. A number of flowering

Remains of Late Cretaceous plants. (Top) The ginkgophyte *Ginkgo wintonensis*. (Bottom) Cone and foliage fragments of the conifer *Austrosequoia wintonensis*. Both specimens are from the Albian-Cenomanian Winton Formation of central Queensland.

plants have also been recorded; magnolias have been identified from leaf impressions, mineralised flowers (*Lovellea wintonensis*) and fruiting structures. Similar Cenomanian macrofloras (with spores and pollen) are known from localities in the Gippsland and Otway basins of southern Victoria. These show some regional differences including a lower diversity of angiosperms, and a generally more pervasive wet-zone adapted flora.

Other Late Cretaceous plants are known almost entirely from microfossil evidence, although some cuticle and leaf fragments have been recovered from drillcores in the Otway Basin. These indicate profound vegetational changes around the Cenomanian/Turonian boundary, with the introduction of modern elements such as holly (*Ilex*) and proteas (e.g. *Banksia*). These probably entered Australia via southern high-latitude migration routes, and by the Santonian had become established elements of the flora. By the Campanian-Maastrichtian, modern proteacean lineages (e.g. *Macadamia*) had started to differentiate, together with other characteristic austral angiosperms such as the Winteraceae and southern beech (*Nothofagus*). The southern beech seems to have originated somewhere in southern South America/Antarctic peninsula and today is distributed throughout many of the ancient Gondwana continental fragments, including Australia, New Guinea, South America, New Zealand and New Caledonia.

The reproductive structures of conifers have been found in Albian-Cenomanian beach sediments of the Mackunda Formation in central Queensland. This cone is from a taxodiacean pine. Image courtesy David and Judy Elliott.

The Cenomanian freshwater bivalve *Pledgia eyrensis*. This genus was named after the palaeontologist Neville Pledge (South Australian Museum), and is known only from the Winton Formation.

Marine and non-marine invertebrates

Knowledge of Australian Late Cretaceous macroinvertebrates is substantial, with several well-preserved assemblages yielding comprehensive data on overall faunal diversity and relationships. The vast majority of specimens are derived from marine rocks, although a few Cenomanian freshwater bivalves (*Pledgia*, *Hyridella*, *Megalovirgus*, *Alathyria*), gastropods (*Melanoides*) and insects (the dragonfly *Aeschnidopsis flindersiensis* and a mecopteran fly) have been identified from the Winton Formation.

Cenomanian marine macroinvertebrates are best-known from the Bathurst Island Group (Walangarlu Mudstone and Moonkinu Formation), which records ~25 genera and species of molluscs including bivalves, gastropods, scaphopods and cephalopods. Several common forms (e.g. bivalves

Wing of the dragonfly *Aeschnidopsis flindersiensis* (with outline lower left) from the Albian-Cenomanian Winton Formation, near Winton in central Queensland. Specimens of *Aeschnidopsis* have also been found the Lower Cretaceous Toolebuc Formation of Queensland. Image courtesy David and Judy Elliott.

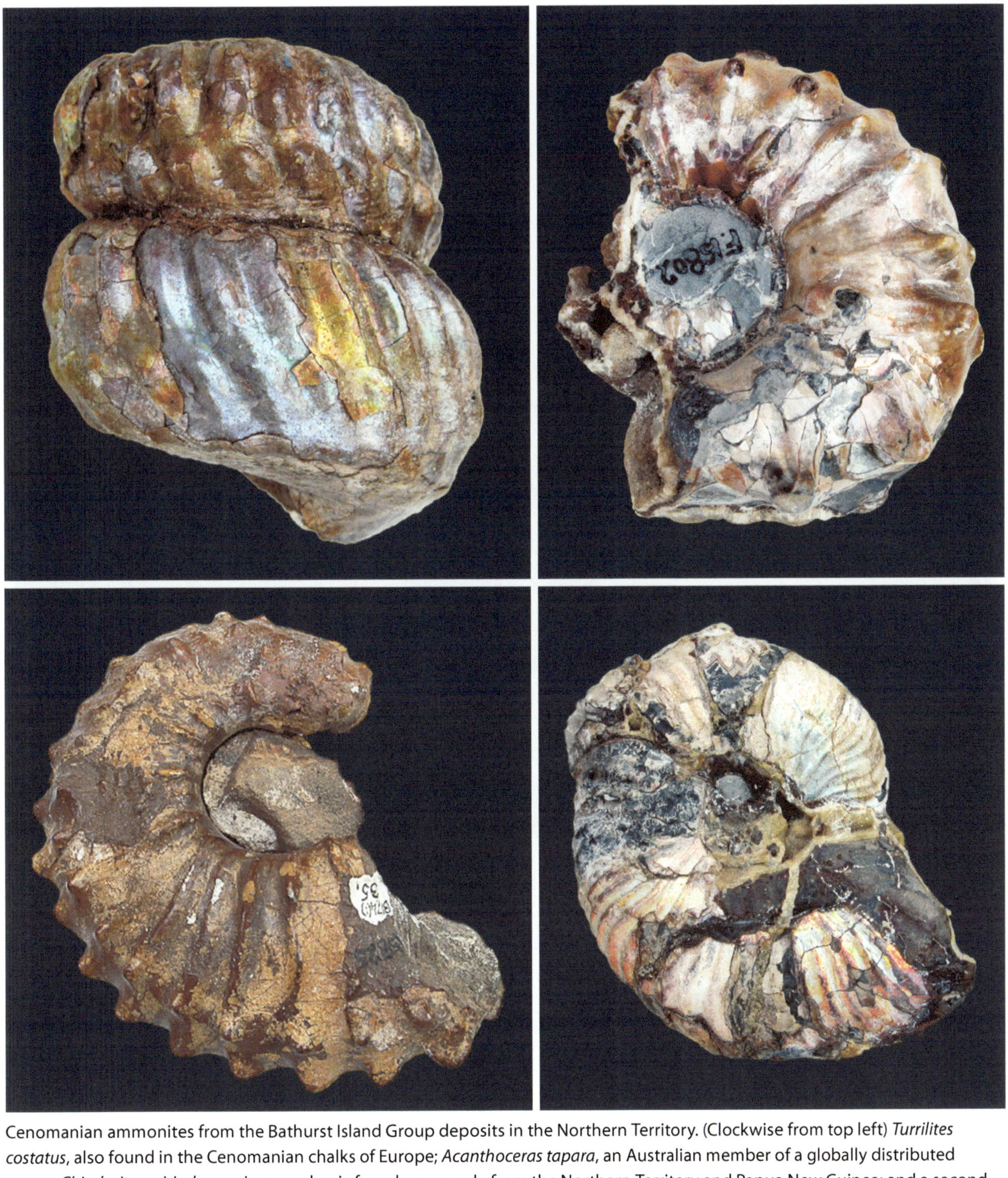

Cenomanian ammonites from the Bathurst Island Group deposits in the Northern Territory. (Clockwise from top left) *Turrilites costatus*, also found in the Cenomanian chalks of Europe; *Acanthoceras tapara*, an Australian member of a globally distributed genus; *Chimbuites mirindowensis*, an endemic form known only from the Northern Territory and Papua New Guinea; and a second specimen of *Acanthoceras tapara,* showing the distinctive spinose ornamentation.

Panopea, *Nuculana* and *Inoceramus*) also occur in Lower Cretaceous deposits. Some new taxa make their first appearance – the bivalve *Nemocardium* and gastropod *Arrhoges* are the oldest records for their groups. The Bathurst Island Group ammonite assemblages are rich; they include a number of globally distributed genera including *Acanthoceras*, *Euomphaloceras*, *Hamites*, *Hypoturrilites*, *Scaphites*, *Sciponoceras* and *Turrilites*. This probably reflects open sea connections – some northern Australian species have also been found in North Africa, Madagascar (e.g. *Scaphites dailyi*) and Europe (e.g. *Turrilites costatus*).

Turonian and Coniacian assemblages are poorly documented because of limited rock outcrop but include ammonites such as *Collingnoniceras*. This genus has been found in the Turonian sediments of the Moonkinu Formation, and is widespread throughout the northern hemisphere. Turonian-Coniacian benthic invertebrates are virtually unknown. Sporadic occurrences in the Molecap Greensand and the Gingin Chalk record up to eight bivalve genera including the widespread *Gryphaea* and *Inoceramus* (well-known from the northern hemisphere), the oyster *Ostrea* (a bivalve that is still extant) and possibly the earliest specimens of the modern scallop *Spondylus*. Other invertebrates include sponges (*Porosphaera*), brachiopods (*Boucharidella*, *Kingena* and *Inopinarticula*), ammonites (e.g. *Baculites*), sea urchins (*Cidaris*) and the stalkless crinoids *Uintacrinus* and *Marsupites*; the latter provide age ranges for the Gingin Chalk through correlations with Europe and North America. Despite their cosmopolitan nature, the Molecap Greensand and Gingin Chalk assemblages herald the appearance of distinctive southern high-latitude kossmaticeratid ammonites such as *Kossmaticeras*. This may indicate the onset of cooler temperatures, a trend

Miria Formation bivalves and gastropods. (Left) The spondylid scallop *Spondylus schekkermanae*. (Right) The cosmopolitan pleurotomariid gastropod *Conotomaria*. The Australian species *C. minacis* has extended the range of *Conotomaria* through to the Maastrichtian.

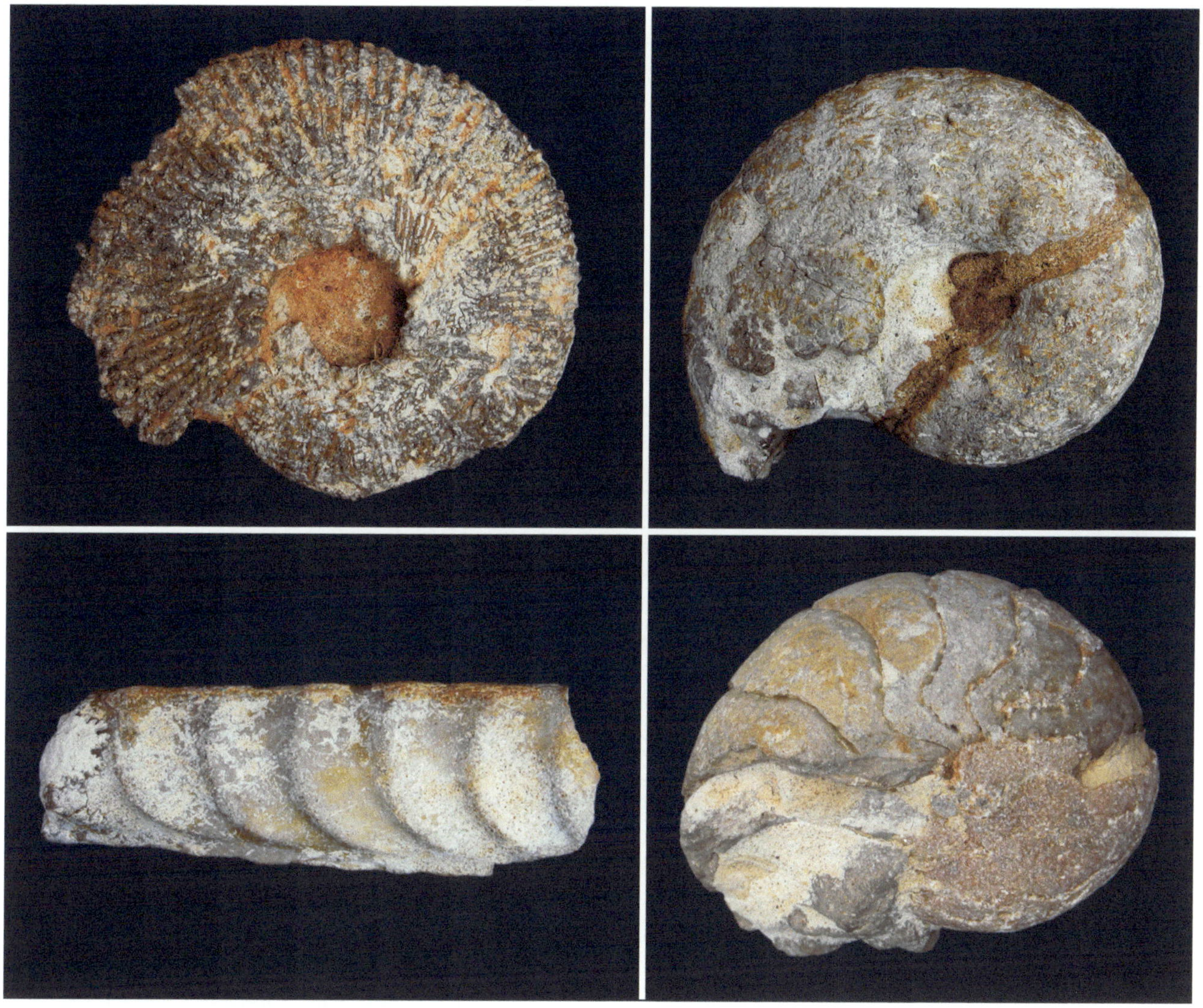

A selection of Miria Formation cephalopods. (Clockwise from top left) The kossmaticeratid ammonite *Gunnerites kalika*, cosmopolitan pachydiscid *Pachydiscus jacquoti*, a nautiloid *Cimomia tenuicostata* and part of the straight-shelled ammonite *Eubaculites carinatus*.

that persists into the latest part of the Cretaceous at least in the southern Perth Basin.

Maastrichtian macroinvertebrate assemblages are some of the richest from the Australian continent. In comparison, the Campanian record is virtually non-existent. The upper Maastrichtian Miria Formation has yielded over 44 documented bottom-dwelling mollusc genera (including gastropods, bivalves and scaphopods), almost all of which are also reported from the Tethyan Oceanic Realm to the north. This is thought to reflect a predominance of warmer waters. The Miria Formation contains a high proportion of infaunal or burrowing bivalves (e.g. *Trigonia* and *Panopea*) and free-swimming pectinids (scallops such as *Chlamys* and *Giraliapecten*), which probably indicate shallow water conditions.

The Miria Formation ammonites are incredibly diverse (~30 genera and 88 species) and include a mixture of southern high-latitude (e.g. kossmaticeratids such as *Gunnerites*) and northern warm-water elements (e.g. *Pachydiscus* and the straight-shelled *Eubaculites* and *Baculites*). The latter appear to derive from contemporary assemblages in south-eastern Africa, particularly southern India. With the occurrence of other cosmopolitan cephalopods such as the nautiloid *Cimomia*, they reinforce the view that Australian latest Cretaceous marine faunas were latitudinally transitional, comprising groups from southern cool- and northern warm-water regions.

Sharks and fish

Considerable work has been undertaken on shark faunas from the Upper Cretaceous deposits of Western Australia, particularly the middle Cenomanian Gearle Siltstone and Alinga Formation of the Murchison River area in the southern Carnarvon Basin. Remains from the Gearle Siltstone include the white pointer-like *Dwardius woodwardi* (also known from chalk deposits in England) and *Cardarbiadon ricki*. *Cardarbiadon* is represented by vertebrae nearly 90 mm in diameter – a size ratio comparable to the largest dimensions of modern white pointer sharks, which can reach lengths of over 5 m. Shark fossils from the Alinga Formation comprise widespread and long-ranging species, most notably *Cretalamna appendiculata* and *Cretoxyrhina mantelli*. Other taxa include *Anomotodon* and *Johnlongia*, both of which have good records from the Cretaceous in Europe and North America.

Turonian-Coniacian shark assemblages from the Molecap Greensand and Gingin Chalk are comparatively poorly documented, but isolated teeth and large vertebral centra attributable to indeterminate lamniform sharks have been recovered. The principal finds include:

- the hexanchoid (cow shark) *Notorynchus aptiensis*;
- the squalids (dog fish) *Centrophoroides* and *Protosqualus*;
- the mitsukurinid-like *Anomotodon*, well-known from Europe and North America;
- the long-ranging *Squatina*;
- the lamnids *Protolamna*, *Cretalamna* (*C. appendiculata* and the endemic species *C. gunsoni*) and *Carcharias*, the last being the most common shark genus in the Molecap Greensand.

The anacoracid *Squalicorax kaupi* and palaeospinacid *Paraorthacodus* have been reported from the Gingin Chalk. Chimaeroid, or rat-fish, remains may include a species of the globally distributed *Ischyodus*. A mandibular toothplate attributable to the chimaerid *Edaphodon* (found elsewhere in the Early Cretaceous of the Lake Eyre region, South Australia) has also been found.

Campanian-Maastrichtian shark fossils are virtually unknown in Australia. Teeth of the lamnid *Enchorhinus eyrensis* have been reported from dredged Maastrichtian rocks in the Great Australian Bight and the Miria Formation at Cardabia Station in Western Australia. A large anacoracid shark vertebra was recently collected from Miria Formation deposits in the Giralia Ranges and has been tentatively assigned to the genus *Squalicorax*. This genus had a near-worldwide distribution throughout much of the Late Cretaceous but did not persist past the Mesozoic.

Bony fish remains are rare in the Molecap Greensand and none have been recorded from the Maastrichtian Miria Formation or Korojon Calcarenite. The cosmopolitan Cretaceous swordfish-like pachycormid *Protosphyraena* has been reported from the Gingin Chalk. A few isolated teleost teeth have been discovered, but have not been formally identified or described.

The non-marine Winton Formation has also produced fish remains, including fragmentary teleosts and numerous lungfish. The isolated lungfish fossils can be assigned to *Metaceratodus*, one of the largest Australian lungfish, with toothplates up to 60 mm across. *Metaceratodus* comprises two species: *M. wollastoni*, which occurs elsewhere in the Lower Cretaceous deposits of Lightning Ridge and the Otway Ranges, and the smaller *M. ellioti*, known from the Winton Formation in Queensland (Winton region) and South Australia (Lake Eyre).

Turtles

Very little diagnostic material is available for Australian Late Cretaceous turtles. Fragmentary remains, including shell parts and limb bones, of a dinnerplate-sized form have been recovered from the Albian-Cenomanian Winton Formation. Preliminary assessment suggests that these fossils might represent a chelid, the group that incorporates all the native freshwater turtles currently living in Australia. If correct, this would have significant palaeobiogeographical implications because chelids of similar age are known from Patagonia, perhaps suggesting that these turtles had a trans-Antarctic distribution during the Cretaceous.

Reconstruction of the ammonite *Pachydiscus*. This taxon is extremely common in the Maastrichtian Miria Formation.

Fish remains from the Cenomanian-Coniacian marine deposits of Western Australia. (Left to right) Shark teeth from the Gearle Siltstone – *Cretalamna gunsoni* and *Dwardius woodwardi*, a lamniform shark vertebra from the Molecap Greensand and a jaw fragment with characteristic lance-shaped tooth (cast) of the pachycormid fish *Protosphyraena* from the Gingin Chalk. Images courtesy Mikael Siversson and the Western Australian Museum.

Plesiosaur and ichthyosaur specimens from the Molecap Greensand. (Left to right) Two views of an isolated elasmosaurid dorsal vertebra thought to show similarities to *Mauisaurus*, and two associated ichthyosaur vertebrae from the basal conglomeritic layer of the Molecap Greensand.

A single incomplete scapula (part of the shoulder girdle) of a large turtle (~2 m long) from the Maastrichtian Miria Formation constitutes the only evidence of Late Cretaceous sea turtles or chelonioids from the Australian continent. Another potential marine turtle specimen (a weathered carapace) has been reported from the Winton Formation, but the identification has yet to be verified.

Ichthyosaurs

Late Cretaceous ichthyosaur fossils are rare in Australia and are restricted to the Cenomanian. Previous reports of ichthyosaur vertebrae from the Maastrichtian Miria Formation have been reinterpreted as belonging to lamniform sharks. Both the Molecap Greensand and Alinga Formation have produced remains attributable to the widespread Cretaceous genus *Platypterygius*. Derivation of the Molecap Greensand material, found near Dandaragan, is contentious since it occurs in the same formation as advanced mosasaur remains (*Platecarpus*, discussed below). This is significant because ichthyosaurs are generally thought to have become extinct by the end of the Cenomanian, long before the first advanced mosasaurs appeared. Importantly, the Molecap Greensand ichthyosaur specimen derived from reworked sediments near the base of the unit, that were probably eroded and redeposited from older strata.

Plesiosaurs

Australian Late Cretaceous plesiosaur remains are known from isolated vertebrae and teeth attributable to elasmosaurids. One large dorsal vertebra from the Molecap Greensand shows similar morphology to those of *Mauisaurus*, a Campanian-Maastrichtian elasmosaurid whose bones have been found in New Zealand, Antarctica and southern South America (Patagonia). If this comparison proves meaningful, the Australian specimen supports a southern high-latitude distribution for *Mauisaurus* and extends the age range for that genus to at least the Coniacian.

Interestingly, a small tail vertebra from the Molecap Greensand has been described as closely comparable to those of the Early Cretaceous Australian pliosauroid '*Leptocleidus*' *clemai*. Diagnostic remains of this taxon have not been recorded from sediments younger than Hauterivian-Barremian and the genus itself does not extend beyond the Lower Cretaceous. The general structure of this vertebra is not dissimilar from that of polycotylids. Consequently, identification of the specimen should be treated with caution until more diagnostic material is recovered.

Marine lizards (mosasaurs and dolichosaurs)

The Australian Late Cretaceous lizard record is currently limited to mosasaurs (Mosasauridae) and dolichosaurs. These long-bodied aquatic predators were closely related to snakes and monitors, and

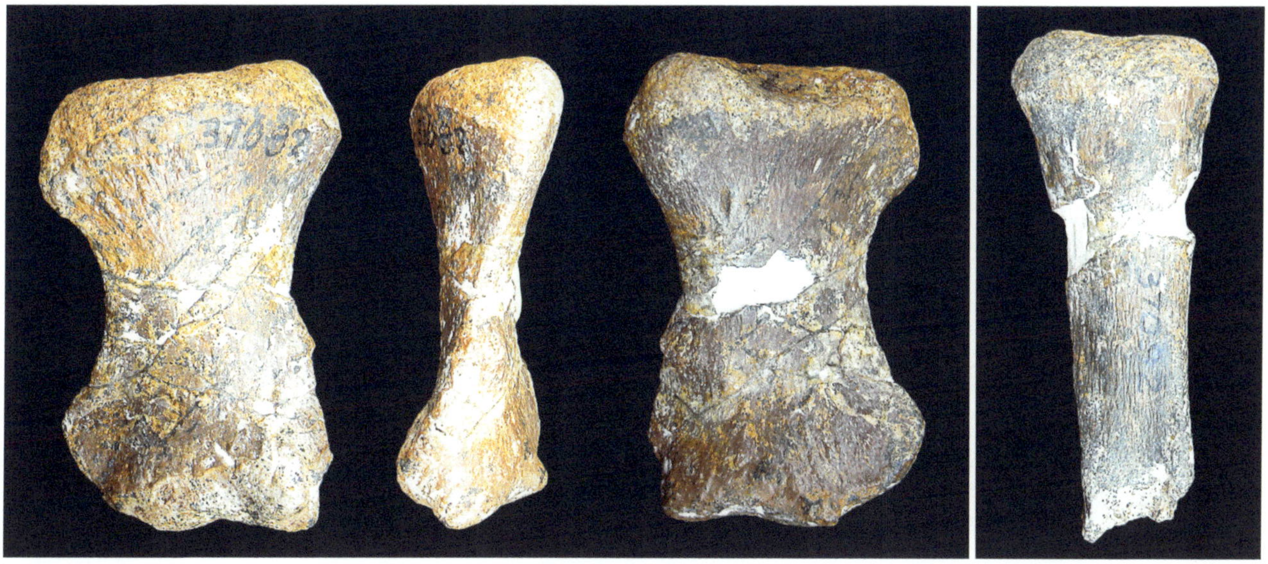

Mosasaur limb elements from the Molecap Greensand, attributed to the widely distributed plioplatecarpine *Platecarpus*. (Left to right) An ulna in ventral, anterior and dorsal views, and an associated incomplete phalanx (finger bone). The slender shape of the phalanx is diagnostic for *Platecarpus*.

were the dominant group of Mesozoic marine reptiles in Late Cretaceous faunas around the world.

The Australian dolichosaur remains are represented by a single vertebra just over 5 mm long, found in the Albian-Cenomanian Winton Formation deposits near Winton in Queensland. This enigmatic specimen is notable because dolichosaurs were typically marine although, as evidenced by this find, they did occasionally invade freshwater environments.

Australian mosasaurs were first recognised from a fragmentary forelimb (ulna and phalanx) from the Molecap Greensand of Western Australia. The slender proportions of these bones suggests possible affinity with *Platecarpus*. This small-bodied (~3 m) plioplatecarpine mosasaur was very widely distributed, occurring in deposits from the Arctic of North America south to Antarctica. The Molecap Greensand has yielded other mosasaur remains, such as a series of ten articulated trunk/tail vertebrae from a locality north of Dandaragan. This material is being studied but appears to belong to a moderately sized plioplatecarpine.

Maastrichtian mosasaur remains have been recovered from the Korojon Calcarenite, a unit stratigraphically underlying the Miria Formation. They include an isolated mosasaur dorsal vertebra.

Reconstruction of *Platecarpus*, a small mosasaur ~3 m long.

A partially prepared skull of the small (~1 m long) crocodile *Isisfordia duncani*, with the lower jaw still in articulation.

The Miria Formation has also yielded sporadic material, with three associated trunk/tail vertebrae from Cardabia Station in the Giralia Ranges. These specimens were particularly well-preserved and appear to represent a large (~6–8 m) mosasaurid of uncertain affinity. Numerous other fragmentary mosasaur vertebrae have come from the Miria Formation. They are taxonomically indeterminate but suggest the presence of a small-bodied probable plioplatecarpine in the deposit.

Crocodiles

Australian Upper Cretaceous crocodile material is generally scarce, comprising isolated teeth, limb bones, vertebrae and scutes. It is known only from the Albian-Cenomanian Winton Formation. A single well-preserved partial skeleton, attributed to the small-bodied (~1 m long) *Isisfordia duncani*, has been described from Winton Formation deposits near Isisford in central-western Queensland. Unusually, *Isisfordia* has reduced back armour and an advanced

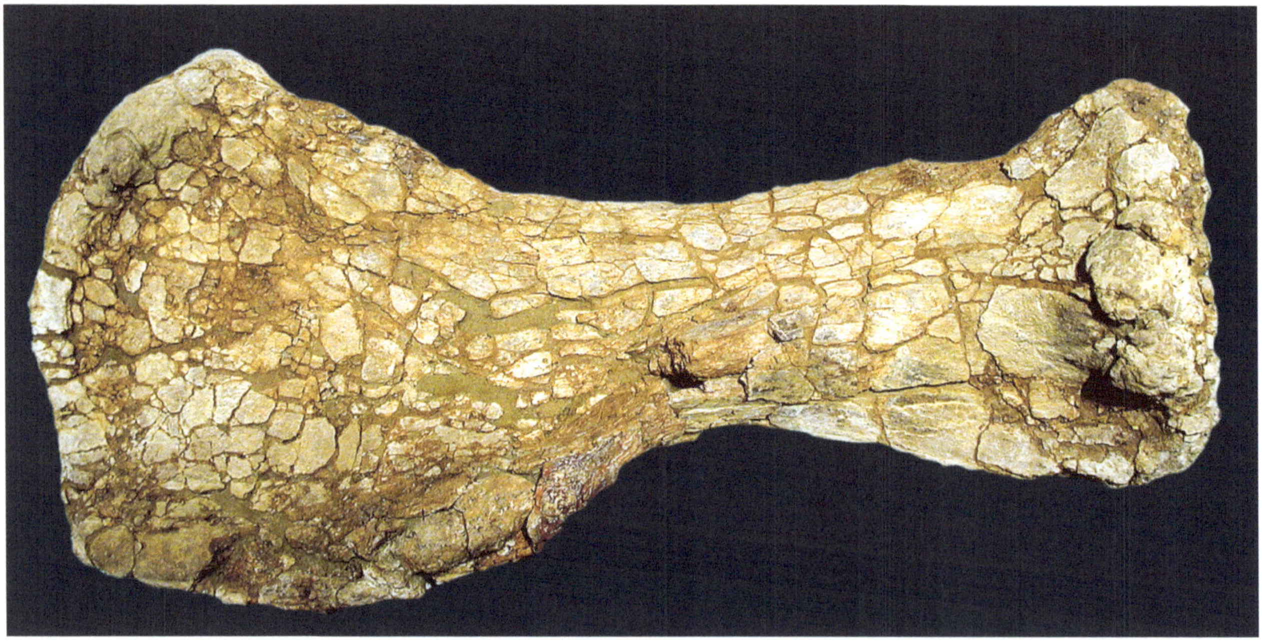

The 1.5 m long humerus of a sauropod from the Albian-Cenomanian Winton Formation deposits near Eromanga in Queensland. This recently discovered specimen is estimated to have come from an animal close to 25 m in length.

structure in the vertebrae and skull, suggesting increased lateral flexibility for efficient swimming and structural reinforcement of the cranium to tackle larger prey. These distinctive features place *Isisfordia* as one of the most primitive members of the modern crocodylian group, Eusuchia, which includes gavials, alligators and crocodiles. Advanced crocodilians thus appear to have undertaken at least part of their early evolution in the Gondwanan region, with their later successful radiation perhaps prompted by a major shift in feeding/locomotory behaviour.

Dinosaurs

Australian Late Cretaceous dinosaurs have gained attention through the growing collection of material from the Winton Formation in Queensland. Sauropod remains are most common, with numerous vertebrae, limb and girdle bones attributable to titanosauriforms, the dominant group of Late

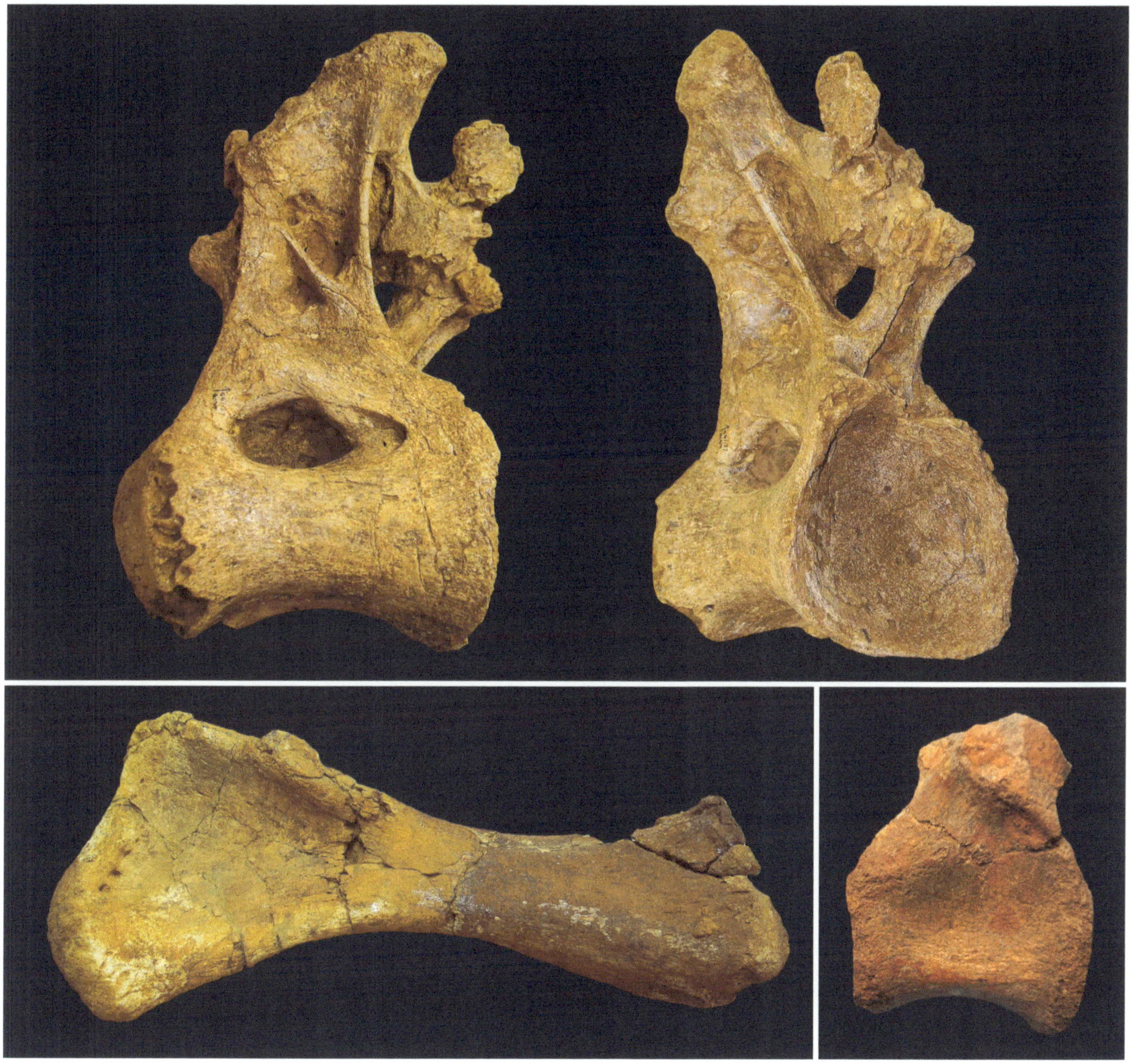

Dinosaur remains from the Albian-Cenomanian Winton Formation of Queensland. (Clockwise from top) A massive sauropod dorsal vertebra in lateral and anterior views, an ornithopod dorsal vertebra, and the humerus of *Diamantinasaurs matildae* in lateral view. Image courtesy David and Judy Elliott.

Bones of *Australovenator wintonensis* from the Albian-Cenomanian Winton Formation of Queensland. A massive hooked claw and phalanx from the hand (top), a pedal phalanx, and the dentary part of the lower jaw (bottom). Images courtesy David and Judy Elliott.

Cretaceous sauropods from Gondwanan continents. At least two new taxa have been identified: *Wintonotitan wattsi*, a primitive titanosaurian of uncertain affinities, and *Diamantinasaurus matildae*, a more advanced form possibly related to saltasaurids – bizarre armoured sauropods best-known from the Cretaceous of South America. Both *Wintonotitan* and *Diamantinasaurus* were huge animals up to 25 m long. Together with reports of a possible third, as yet undescribed, sauropod taxon, they indicate that a surprisingly diverse assemblage of mega-herbivores occupied the lowland floodplains of central Queensland ~100 million years ago.

Other dinosaur body fossils from the Winton Formation include small and large ornithopods, which are known only from isolated teeth, phalanges, limb bones and vertebrae. Theropods are also rare, with a handful of teeth and one partial skeleton discovered near Winton in central Queensland. Named *Australovenator wintonensis*, that specimen is Australia's most complete carnivorous dinosaur. It comprises elements from the fore- and hindlimbs, part of the pelvis, some ribs, the lower jaw and a number of loose teeth. At ~5 m in length, *Australovenator* was a medium-sized, lightly built and probably fast-running theropod with enlarged sickle-shaped claws on its hands. These may have been used to grapple with prey, although association of *Australovenator* with the carcass of a massive sauropod could imply that carrion formed a component of the diet. *Australovenator* appears to be a primitive carcharodontosaurid, a predominantly Gondwanan group of allosauroids that includes some South American and African forms larger than the famous North American *Tyrannosaurus*. Curiously, the nearest relatives of *Australovenator* hail from the northern hemisphere (England and Japan), possibly implying a near-global distribution of allosauroids during the Early-Late Cretaceous.

The Lark Quarry dinosaur trackway during its initial excavation in 1977. The large *Tyrannosauropus* trackway is clearly visible (arrows) in the middle distance. Tracks of the smaller bipedal dinosaurs (*Skartopus* and *Wintopus*) pepper the entire surface of the rock layer. Image courtesy Alex Ritchie.

The famous Winton dinosaur trackway at Lark Quarry south of Winton preserves evidence of large (indeterminate) and small (*Wintopus*) ornithopods, small theropods (*Skartopus*) and a large carnosaur (*Tyrannosauropus*). These footprints have been interpreted as evidence of a dramatic dinosaur stampede – the tracks mostly progress in the same direction and have minimal overlap, suggesting that the smaller dinosaurs were moving as a group. However, the three-toed prints of the large carnivore head the opposite way. They may have been made as the predator stalked the smaller animals, prompting them to flee.

Isolated Upper Cretaceous dinosaur fossils have been found in the marine Molecap Greensand and Miria Formation; these probably represent the remains of carcasses washed out to sea. Specimens include an allosauroid-like theropod toe bone from the Molecap Greensand at Molecap Hill Quarry, near Gingin. Material from the Miria Formation includes a large (210 mm long) humerus, recovered from the Giralia Ranges, that

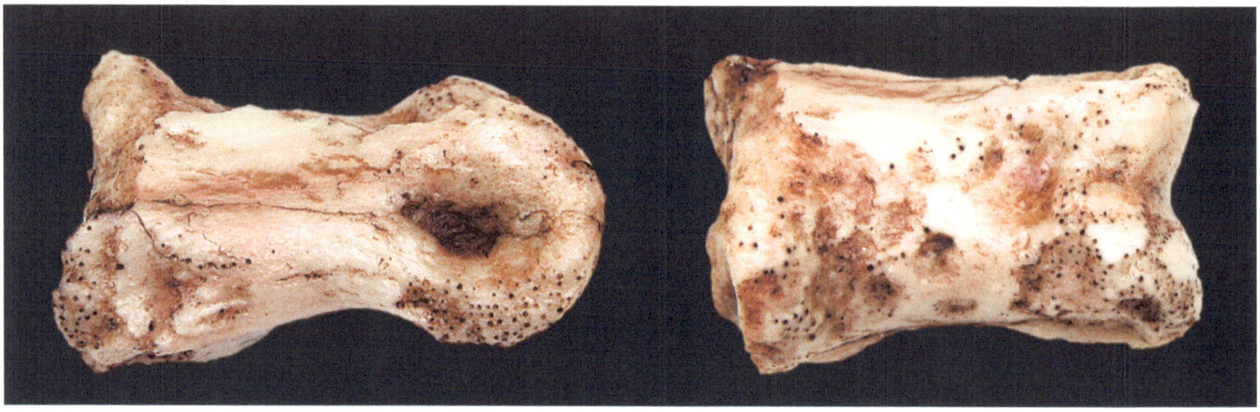

Theropod phalanx (cast) from the Molecap Greensand of Western Australia.

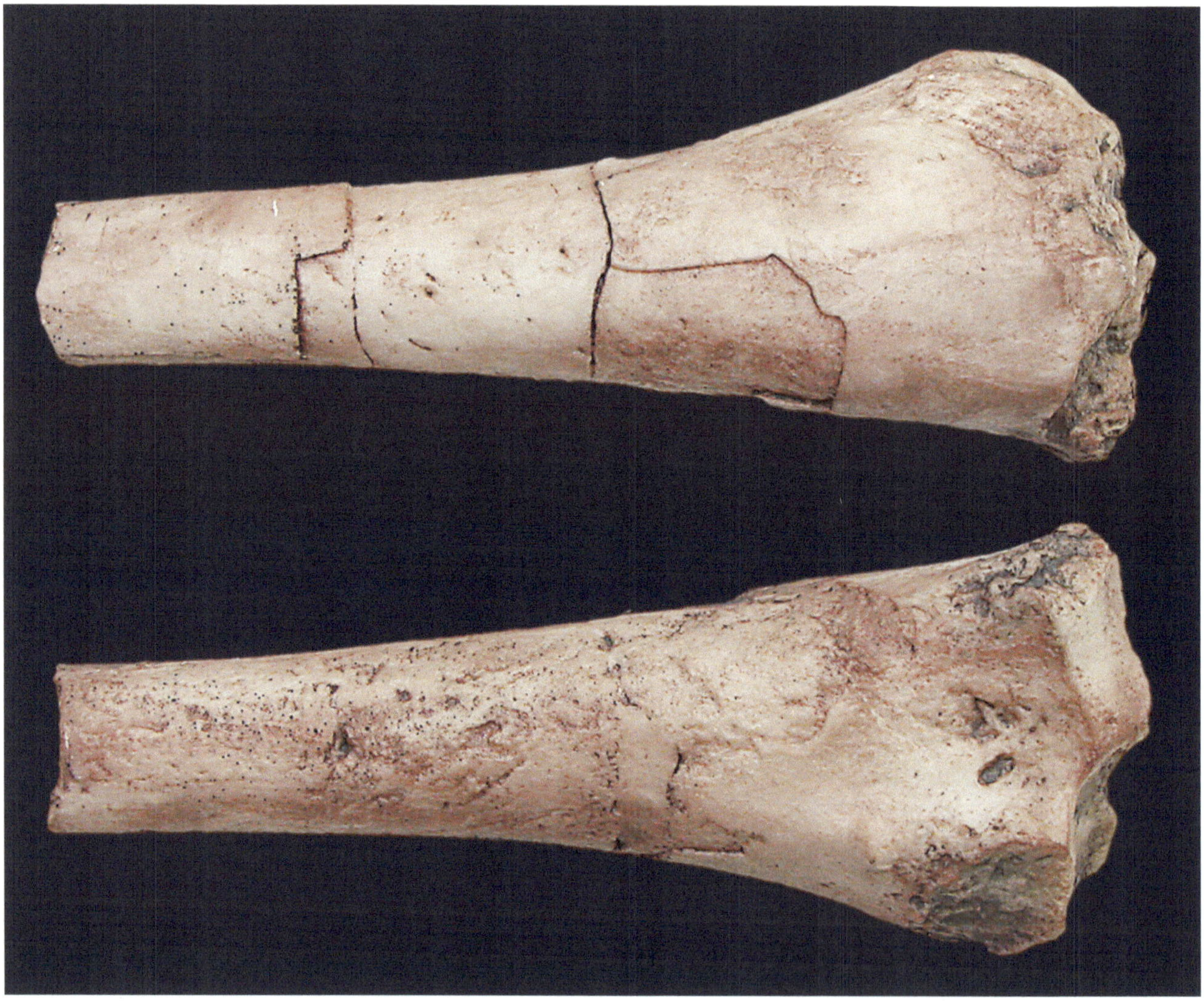

Isolated azhdarchid pterosaur ulna (cast) from the Maastrichtian Miria Formation, Western Australia. The bone is shown in dorsal (top) and ventral (bottom) views. The Miria Formation pterosaur specimen is one of only a handful known from the Australian continent, and suggests that these giant flying reptiles were present in Cretaceous high-latitude regions until the very end of the Age of Dinosaurs.

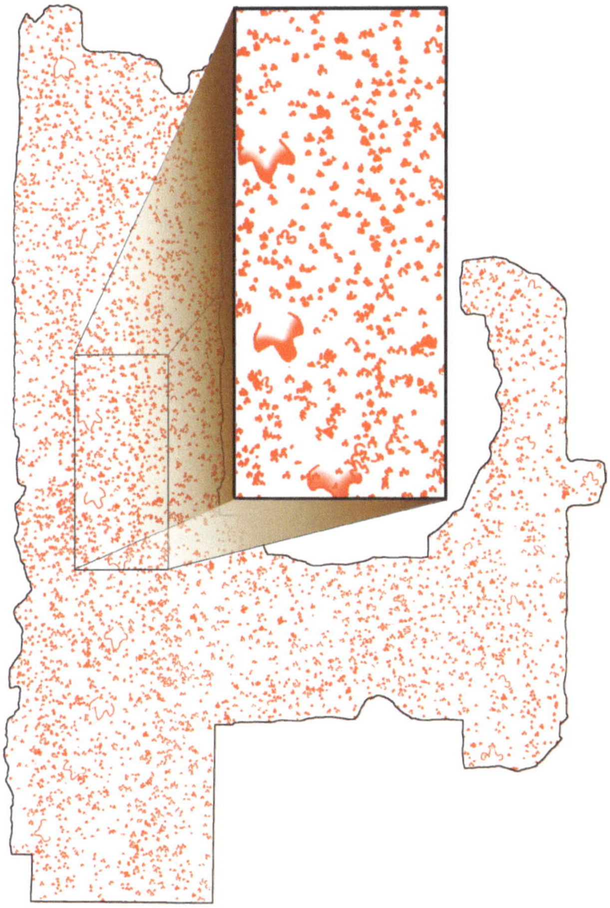
A schematic map of the exposed Winton dinosaur trackway (after Thulborn and Wade 1984). The footprints (marked in red) have been interpreted as evidence of a dinosaur stampede.

shows similarities to the humerus of tetanuran theropods and dromaeosaurs.

Pterosaurs

Late Cretaceous pterosaur remains are exceedingly rare in Australia. The only known occurrences are an incomplete limb bone from the Winton Formation of Queensland, part of a jaw from the Molecap Greensand of Western Australia, and a single ulna from the Miria Formation deposits of Toothawarra Creek, in the Giralia Ranges of Western Australia. The last represents the only diagnostic element, and suggests a pterosaur with a 5 m wingspan. The general structure of the ulna is very similar to those of azhdarchids, particularly the poorly known Maastrichtian genus *Arambourgiana* from Jordan. The Cretaceous azhdarchids include some of the largest pterosaurs of all time. With a wingspan of over 11 m, *Quetzalcoatlus* is a famous giant from the Late Cretaceous of North America.

GLOSSARY

benthic living on the bottom of sea, lake, river etc.

biogeography aspect of biology dealing with the geographical distribution of animals and plants

bioturbation the disruption of sediments by the activities of burrowing organisms

carapace bony plates (covered by horny tissue) making up the dorsal shell of turtles and tortoises

endemi/c (ism) restricted to a certain geographic region

epicontinental located upon a continental plateau or platform

epifaunal living on the surface of the ocean floor (e.g. an encrusting fauna)

facies any observable attribute of a rock or stratigraphic unit (e.g. composition, overall appearance)

family taxonomic group of closely related genera; in the scientific literature, family names formally begin with a capital letter

femur thigh bone, the largest bone of the hindlimb in vertebrates

fibula bone positioned outside or posterior to the tibia in the shin of vertebrates

fluvial resulting from river action

genus (*plural* genera) taxonomic group of closely related species; in the scientific literature, generic names are italicised

glauconite a clay mineral characteristically occurring in marine sediments

glendonite(s) crystal pseudomorph structure of the calcium hexa-hydrate mineral ikaite; glendonites form only in marine sediments between –1°C and 6°C and are commonly used as palaeoclimatic indicators for very cold water temperatures

humerus upper bone of the forelimb of vertebrates

ichnofossil/ichnotaxon a trail, track or burrow made by an animal and preserved in rock/name given to a ichnofossil (also termed tracefossil); in the scientific literature, ichnotaxon names are italicised

ilium dorsal bone of the pelvic girdle

lacustrine resulting from or produced by lakes

lithology description of the physical character of a rock (e.g. colour, structure, mineralogical properties)

macrofossil fossils that are visible without magnification

macroinvertebrate invertebrates that are visible without magnification

metacarpal(s) bones joining the wrist to fingers

metatarsal(s) bones joining the ankle to toes

microfossil fossils which are not visible without magnification

orbit bony cavity in which the eye is situated

ossicles any small bone or calcified hard structure

pectoral girdle skeletal support in the shoulder region of vertebrates for attachment of the forelimb; made up of several bones – the clavicle (occasionally plus interclavicle), scapula and coracoid

pelagic living at middle or suface levels of the ocean

pelvic girdle skeletal support in the hip region of vertebrates for attachment of the hindlimb; made up of several bones – the ilium, ischium and pubis

phalanx one of several bones (phalanges) in fingers or toes

phosphorite a sedimentary rock composed principally of phosphate minerals

plastron bony plates (covered by horny tissue) making up the underside of the shell in turtles and tortoises

quadratojugal dermal bone connecting the quadrate and jugal bones in the rear section of the skull

scapulocoracoid bone made up of the fused scapula and coracoid in the pectoral girdle of pterosaurs

scute an external scale or bone present within the skin

steinkerns rock structure formed from consolidated sediment that has filled a hollow organic structure (e.g. shell); often preserved after the dissolution of the mould (also termed an endocast)

taxon (*plural* taxa) members of any particular taxonomic group (e.g. species, genus, family)

taxonomy analysis of an organism's characteristics for the purposes of classification

terrestrial living or found on land

tetrapod a four-footed animal – amphibians, reptiles, birds and mammals

tibia larger bone positioned between the knee and ankle, and inside the fibula in the shin of vertebrates

tillite sedimentary rock formed from till – unstratified sediments comprising a mixture of clay, sand, gravel and boulders accummulated and deposited by glacial movement

ulna long bone of the forearm, parallel to the radius in vertebrates

vertebral centrum main body of a vertebra, from which the neural and haemal arches arise

BIBLIOGRAPHY

The bibliography contains the primary reference works consulted for the various sections of this book, plus a selection of recent publications (over the last 20 years) on Australia's Mesozoic fossil deposits and associated floras and faunas.

Significant Australian Triassic deposits and localities

Conaghan PJ 1980. The Hawkesbury Sandstone: gross characteristics and depositional environment. *Geological Survey of New South Wales Bulletin* **26**, 189–253.

Herbert C 1980. Winamatta Group and Mittagong Formation. *Geological Survey of New South Wales Bulletin* **26**, 255–272.

Herbert C 1997. Sequence stratigraphic analysis of early and Middle Triassic alluvial and estuarine facies in the Sydney Basin, Australia. *Australian Journal of Earth Sciences* **44**, 125–143.

Jensen AR 1975. Permo-Triassic stratigraphy and sedimentation in the Bowen Basin, Queensland. *Bureau of Mineral Resources, Geology and Geophysics Bulletin* **154**, 1–187.

Kwitko G 1995. Triassic intramontane basins. In *The Geology of South Australia. Volume 2. The Phanerozoic.* (Eds Drexel JF and Preiss WV) pp. 98–101. Geological Survey of South Australia, Bulletin 54.

McDonnell KL 1974. Depositional environments of the Triassic Gosford Formation, Sydney Basin. *Journal of the Geological Society of Australia* **21**, 107–132.

McDonnell KL 1980. Notes on the depositional environments of the Terrigal Formation. *Geological Survey of New South Wales Bulletin* **26**, 171–176.

McKellar JL 2004. Geophysical controls on late Palaeozoic-early Mesozoic geological history and floral succession: eastern Australia in perspective. *Memoir of the Association of Australasian Palaeontologists* **29**, 47–83.

McWae JRH, Playford PE, Lindner AW, Glenister BF and Balme BE 1958. The stratigraphy of Western Australia. *Journal of the Geological Society of Australia* **4**, 1–161.

Ward CR 1980. Notes on the Bulgo Sandstone and the Bald Hill Claystone. *Geological Survey of New South Wales Bulletin* **26**, 179–186.

Triassic land plants

Balme BE, Kershaw AP and Webb JA 1995. Floras of Australian coal measures, with notes on their associated Mesozoic faunas. In *Geology of Australian Coal Basins.* (Eds Ward CR, Harrington HJ, Mallett CW and Beeston JW) pp. 41–62. Geological Society of Australia, Coal Geology Group Special Publication 1.

Burges NA 1935. Additions to our knowledge of the flora of the Narrabeen Stage of the Hawkesbury Series in New South Wales. *Proceedings of the Linnean Society of New South Wales* **60**, 257–264.

Cantrill DJ and Webb JA 1998. Permineralised pleuromeid lycopsid remains from the Early Triassic Arcadia Formation, Queensland, Australia. *Review of Palaeobotany and Palynology* **102**, 189–211.

Retallack GJ 1975. Life and times of a Triassic lycopod. *Alcheringa* **1**, 3–29.

Retallack GJ 1977. Triassic palaeosols in the upper Narrabeen Group of New South Wales. Part II: Classification and reconstruction. *Journal of the Geological Society of Australia* **24**, 19–36.

Retallack GJ 1980. Late Carboniferous to Middle Triassic megafossil floras from the Sydney Basin. *New South Wales Geological Survey Bulletin* **26**, 384–430.

Walkom AB 1925. Fossil plants from the Narrabeen stage of the Hawkesbury Series. *Proceedings of the Linnean Society of New South Wales* **1**, 214–224.

White ME 1986. *The Greening of Gondwana.* Reed Books, Sydney.

Triassic marine and non-marine invertebrates

Dickins JM and McTavish RA 1963. Lower Triassic marine fossils from the Beagle Ridge (BMR 10) Bore, Perth Basin, Western Australia. *Journal of the Geological Society of Australia* **10**, 123–140.

Etheridge R Jr 1888. Invertebrate fauna of the Hawkesbury-Winamatta series of New South Wales. *Memoirs of the Geological Survey of New South Wales, Palaeontology* **1**, 1–26.

Fleming PJG 1966. Eotriassic marine bivalves from the Maryborough Basin, south-east Queensland. *Geological Survey of Western Australia Annual Report* **1963–1964**, 55–57.

Hocknull SA 2000. Mesozoic freshwater and estuarine bivalves from Australia. *Memoirs of the Queensland Museum* **45**, 405–426.

Jell PA 2004. Fossil insects of Australia. *Memoirs of the Queensland Museum* **50**, 1–123.

Jell PA and Darragh TA 1998. The fossil record. In *Mollusca: The Southern Synthesis. Fauna of Australia. Vol. 5, Part A.* (Eds Beesley PL, Ross GJB and Wells A) pp. 88–101. CSIRO Publishing, Melbourne.

Ludbrook NH 1961. Mesozoic non-marine Mollusca (Pelecypods, Unionidae) from the north of South Australia. *Transactions of the Royal Society of South Australia* **84**, 139–147.

McMichael DF 1957. A review of fossil freshwater mussels (Pelecypoda: Unionidae) of Australasia. *Proceedings of the Linnean Society of New South Wales* **81**, 222–242.

Riek EF 1950. A fossil mecopteron from the Triassic Beds at Brookvale, NSW. *Records of the Australian Museum* **22**, 254–256.

Riek EF 1954. Further Triassic insects from Brookvale, NSW. *Records of the Australian Museum* **23**, 161–168.

Riek EF 1954. Fossil insects from the Triassic Beds at Mt Crosby, Queensland. *Australian Journal of Zoology* **3**, 654–691.

Runnegar B 1969. A Lower Triassic ammonoid fauna from south-east Queensland. *Journal of Paleontology* **43**, 818–828.

Skwarko SK and Kummel B 1974. Marine Triassic molluscs of Australia and New Guinea. *Bureau of Mineral Resources, Geology and Geophysics Bulletin* **150**, 111–134.

Tillyard RJ 1916. Mesozoic and Tertiary insects of Queensland and New South Wales: descriptions of the fossil insects. *Geological Survey of Queensland Publication* **253**, 11–60.

Tillyard RJ 1917. Mesozoic insects of Queensland. No. 1 Planipennia, Trichoptera, and the new order Protmecoptera. *Proceedings of the Linnean Society of New South Wales* **42**, 175–200.

Tillyard RJ 1919. Mesozoic insects of Queensland. No. 5 Mecoptera, the new order Paratrichoptera and additiona to Planipennia. *Proceedings of the Linnean Society of New South Wales* **44**, 194–212.

Tillyard RJ 1922. Mesozoic insects of Queensland. No. 9. *Proceedings of the Linnean Society of New South Wales* **47**, 447–470.

Tillyard RJ 1923. Mesozoic insects of Queensland. No. 10. *Proceedings of the Linnean Society of New South Wales* **48**, 481–498.

Tillyard RJ and Dunstan B 1923. Mesozoic insects of Queensland. Part I: Introduction and Coleoptera. *Geological Survey of Queensland Publication* **273**, 1–89.

Wilson GDF and Edgecombe GD 2003. The Triassic isopod *Protamphisopus winamattaensis* (Chilton) and comparison with extant taxa (Crustacea, Phraetoicidea). *Journal of Paleontology* **77**, 454–470.

Triassic non-marine fish

Kemp A 1993. Problematic Triassic dipnoans from Australia. In *The Non-Marine Triassic.* (Eds Lucas SG and Morales M) pp. 223–227. New Mexico Museum of Natural History and Science Bulletin 3.

Kemp A 1994. Australian Triassic lungfish skulls. *Journal of Paleontology* **68**, 647–654.

Kemp A 1996. Triassic lungfish from Gondwana. In *Mesozoic Fishes: Systematics and Paleoecology.* (Eds Arratia G and Viohl G) pp. 409–416. Verlag Dr Friedrich Pfeil, München.

Kemp A 1998. Skull structure in post Paleozoic lungfish. *Journal of Vertebrate Paleontology* **18**, 43–63.

Long JA 1991. The long history of Australian fossil fishes. In *Vertebrate Palaeontology of Australasia.* (Eds Vickers-Rich P, Monaghan JM, Baird RF and Rich TH) pp. 337–428. Pioneer Design Studio, Monash University, Melbourne.

López-Arbarello A 2004. The record of Mesozoic fishes from Gondwana (excluding India and Madagascar). In *Mesozoic Fishes 3: Systematic, Paleoenvironments and Biodiversity.* (Eds Arratia G and Tintori A) pp. 597–624. Verlag Dr Friedrich Pfeil, München.

Wade RT 1935. *The Triassic Fishes of Brookvale, New South Wales.* British Museum (Natural History), London.

Wade RT 1940. The Triassic fishes of Gosford, New South Wales. *Journal of the Royal Society of New South Wales* **73**, 206–217.

Wade RT 1941. Australian Triassic fishes. *Journal of the Royal Society of New South Wales* **74**, 377–396.

Wade RT 1953. Note on a Triassic fossil fish from Leigh Creek, South Australia. *Transactions of the Royal Society of South Australia* **76**, 80–81.

Woodward AS 1890. The fossil fishes of the Hawkesbury series at Gosford, New South Wales. *Memoirs of the Geological Survey of New South Wales (Palaeontology)* **1054**, 1–55.

Woodward AS 1908. The fossil fishes of the Hawkesbury series at St Peters. *Memoirs of the Geological Survey of New South Wales (Palaeontology)* **10**, 1–32.

Triassic amphibians

Cosgriff JW 1965. A new genus of Temnospondyli from the Triassic of Western Australia. *Journal of the Royal Society of Western Australia* **48**, 65–95.

Cosgriff JW 1969. *Blinasaurus*, a brachyopid genus from Western Australia and New South Wales. *Journal of the Royal Society of Western Australia* **52**, 69–88.

Cosgriff JW 1973. *Notobrachyops picketii*, a brachyopid from the Ashfield Shale, New South Wales. *Journal of Paleontology* **47**, 1094–1101.

Cosgriff JW 1974. Lower Triassic temnospondyli of Tasmania. *Geological Society of America, Special Paper* **149**, 1–131.

Cosgriff JW 1984. The temnospondyl labyrinthodonts of the earliest Triassic. *Journal of Vertebrate Paleontology* **4**, 30–46.

Cosgriff JW and Defauw SL 1987. A capitosaurid labyrinthodont from the Lower Scythian of Tasmania. *Alcheringa* **11**, 21–41.

Cosgriff JW and Garbutt NK 1972. *Erythrobatrchus noonkanbahensis*, a trematosaurid species from the Blina Shale. *Journal of the Royal Society of Western Australia* **55**, 5–18.

Damiani RJ 1999. Giant temnospondyl amphibians from the Early to Middle Triassic Narrabeen Group of the Sydney Basin, New South Wales, Australia. *Alcheringa* **23**, 87–109.

Damiani RJ 2001. A systematic revision and phylogenetic analysis of Triassic mastodonsauroids (Temnospondyli, Stereospondyli). *Zoological Journal of the Linnean Society* **133**, 379–482.

Damiani RJ and Warren AA 1996. A new look at members of the Brachyopoidea (Amphibia, Temnospondyli) from the Early Triassic of Queensland and a preliminary analysis of brachyopoid relationships. *Alcheringa* **20**, 277–300.

Damiani RJ and Warren AA 1997. Re-interpretation of *Parotosuchus wadei* Cosgriff, a capitosaurid from the Triassic Narrabeen Group at Gosford, New South Wales, with comments on its growth stage. *Alcheringa* **21**, 281–289.

Howie AA 1972. A brachyopid labyrinthodont from the Lower Trias of Queensland. *Proceedings of the Linnean Society of New South Wales* **96**, 268–277.

Nield CM, Damiani R and Warren A 2006. A short-snouted trematosauroid (Tetrapoda: Temnospondyli) from the Lower Triassic of Australia: the oldest trematosaurine. *Alcheringa* **30**, 263–271.

Pepperell J and Grigg G 1974. A labyrinthodont trackway from the Mid-Triassic near Sydney, New South Wales. *Proceedings of the Linnean Society of New South Wales* **99**, 54–56.

Warren AA 1980. *Parotosuchus* from the Early Triassic of Queensland and Western Australia. *Alcheringa* **4**: 25–36.

Warren AA 1981. A horned member of the labyrinthodont superfamily Brachyopoidea from the Early Triassic of Queensland. *Alcheringa* **5**, 273–288.

Warren AA 1985. Two long-snouted temnospondyls (Amphibia, Labyrinthodonta) from the Triassic of Queensland. *Alcheringa* **9**, 293–296.

Warren AA 1985. An Australian plagiosauroid. *Journal of Paleontology* **59**, 236–241.

Warren AA 1991. Australian fossil amphibians. In *Vertebrate Palaeontology of Australasia.* (Eds Vickers-Rich P, Monaghan JM, Baird RF and Rich TH) pp. 569–589. Pioneer Design Studio, Monash University, Melbourne.

Warren AA and Black T 1985. A new rhytidosteid (Amphibia, Labyrinthodonta) from the Early Triassic Arcadia Formation of Queensland, Australia, and a consideration of the relationships of Triassic temnospondyls. *Journal of Vertebrate Paleontology* **5**, 303–327.

Warren AA and Hutchinson MN 1987. The skeleton of a new hornless rhytidosteid (Amphibia, Temnospondyli). *Alcheringa* **11**, 291–302.

Warren AA and Hutchinson MN 1988. A new capitosaurid amphibian from the Early Triassic of Queensland, and the ontogeny of the capitosaur skull. *Palaeontology* **31**, 857–876.

Warren AA and Hutchinson MN 1990. *Lapillopsis*, a new genus of temnospondyl amphibians from the Early Triassic of Queensland. *Alcheringa* **14**, 149–158.

Warren AA and Hutchinson MN 1990. The young ones: small temnospondyls from the Arcadia Formation. *Memoirs of the Queensland Museum* **28**, 103–106.

Warren AA and Marsicano CA 1998. Revision of the Brachiopidae (Temnospondyli) from the Triassic of the Sydney, Carnarvon and Tasmania Basins. *Alcheringa* **22**, 329–342.

Warren AA and Marsicano CA 2000. A phylogeny of the Brachyopidae (Temnospondyli, Stereospondyli). *Journal of Vertebrate Paleontology* **20**, 462–483.

Warren AA, Damiani R and Yates AM 2006. The South African stereospondyl *Lydekkerina huxleyi* (Tetrapoda, Temnospondyli) from the Lower Triassic of Australia. *Geological Magazine* **143**, 877–886.

Watson DMS 1958. A new labyrinthodont (Paracyclotosaurus) from the Upper Trias of New South Wales. *Bulletin of the British Museum of Natural History* **3**, 233–263.

Yates AM 1999. The Lapillopsidae: a new family of small temnospondyls from the Early Triassic of Australia. *Journal of Vertebrate Paleontology* **19**, 302–320.

Yates AM 2000. A new tiny rhytidosteid (Temnospondyli, Stereospondyli) from the Early Triassic of Australia and the possibility of hidden temnospondyl diversity. *Journal of Vertebrate Paleontology* **20**, 484–489.

Yates AM and Warren AA 2000. The phylogeny of the 'higher temnospondyls' (Vertebrata, Choanata). *Zoological Journal of the Linnean Society* **128**, 77–121.

Triassic marine reptiles

Kear BP 2004. Biogeographic and biostratigraphic implications of Australian Mesozoic marine reptiles. *Australian Biologist* **17**, 4–22.

Triassic lizard-like reptiles

Bartholomai A 1979. New lizard-like reptiles from the Early Triassic of Queensland. *Alcheringa* **3**, 225–234.

Molnar RE 1991. Fossil reptiles in Australia. In *Vertebrate Palaeontology of Australasia.* (Eds Vickers-Rich P, Monaghan JM, Baird RF and Rich TH) pp. 605–702. Pioneer Design Studio, Monash University, Melbourne.

Triassic archosaurs

Camp CL and Banks MR 1978. A proterosuchian reptile from the Early Triassic of Tasmania. *Alcheringa* **2**, 143–158.

Kear BP 2009. Proterosuchid archosaur remains from the Early Triassic Bulgo Sandstone of Long Reef, New South Wales. *Alcheringa* **33**, 331–337.

Thulborn RA 1979. A proterosuchian thecodont from the Rewan Formation of Queensland. *Memoirs of the Queensland Museum* **19**, 331–355.

Thulborn RA 1986. The Australian Triassic reptile *Tasmaniosaurus triassicus* (Thecodontia: Proterosuchia). *Journal of Vertebrate Paleontology* **6**, 123–142.

Thulborn RA 1986. Early Triassic tetrapod faunas of south-eastern Gondwana. *Alcheringa* **10**, 297–313.

Triassic dinosaurs

Bartholomai A 1966. Fossil footprints in Queensland. *Australian Natural History* **15**, 147–150.

Long JA 1998. *Dinosaurs of Australia and New Zealand and Other Animals of the Mesozoic Era.* University of New South Wales Press, Sydney.

Thulborn RA 2000. Australia's earliest theropods: footprint evidence in the Ipswich coal measures (Upper Triassc) of Queensland. *Gaia* **15**, 123–142.

Vickers-Rich P, Rich TH, McNamara G and Milner A 1999. Is *Agrosaurus macgillivrayi* Australia's oldest dinosaur? *Records of the Western Australian Museum, Supplement* **57**, 191–200.

Triassic therapsids

Retallack GJ 1996. Early Triassic therapsid footprints from the Sydney Basin, Australia. *Alcheringa* **20**, 301–314.

Thulborn RA 1983. A mammal-like reptile from Australia. *Nature* **303**, 330–331.

Thulborn RA 1990. Mammal-like reptiles in Australia. *Memoirs of the Queensland Museum* **28**, 169.

Significant Australian Jurassic deposits and localities

Brunnschweiller RO 1954. Mesozoic stratigraphy and history of the Canning Desert and Fitzroy Valley, Western Australia. *Journal of the Geological Society of Australia* **1**, 35–54.

Brunnschweiller RO 1957. The geology of the Dampier Peninsula, Western Australia. *Bureau of Mineral Resources, Geology and Geophysics Report* **13**, 1–19.

Burger D 1984. A palynological review of the Jurassic below the Injune Creek Group in the Eromanga Basin, Queensland. *Bureau of Mineral Resources, Australia, Record* **19**, 1–23.

Burger D 1986. Palynology, cyclic sedimentation, and palaeoenvironments in the late Mesozoic of the Eromanga Basin. *Geological Society of Australia, Special Publication* **12**, 53–70.

Burger D 1989. Stratigraphy, palynology and palaeoenvironments of the Hooray Sandstone, eastern Eromanga Basin, Queensland and New South Wales. *Queensland Department of Mines, Report* **3**, 1–28.

Day RW, Whitaker WG, Murray CG, Wilson IH and Grimes KG 1983. Queensland Geology. A companion volume to the 1:2 500 000 scale geological map (1975). *Geological Survey of Queensland Publication* **383**, 1–194.

Embleton BJJ and McElhinny MW 1982. Marine magnetic anomalies, palaeomagnetism and the drift history of Gondwanaland. *Earth and Planetary Science Letters* **58**, 141–150.

Exon NF 1976. Geology of the Surat Basin in Queensland. *Bureau of Mineral Resources, Australia, Bulletin* **166**, 1–160.

Exon NF and Burger D 1981. Sedimentary cycles in the Surat Basin and global changes of sea level. *Bureau of Mineral Resources, Journal of Geology and Geophysics* **6**, 153–159.

McWae JRH, Playford PE, Lindner AW, Glenister BF and Balme BE 1958. The stratigraphy of Western Australia. *Journal of the Geological Society of Australia* **4**, 1–161.

Smart J and Senior BR 1980. Jurassic-Cretaceous basins of north-eastern Australia. In *The Geology and Geophysics of Northern Australia.* (Eds Henderson RA and Stephenson PJ) pp. 315–328. Geological Society of Australia, Queensland Division, Brisbane.

Turner S, Bean LB, Dettmann M, McKellar JL, McLoughlin S and Thulborn T 2009. Australian Jurassic sedimentary and fossil successions: current work and future prospects for marine and non-marine correlation. *GFF* **131**, 49–70.

Jurassic land plants

Douglas JG 1969. The Mesozoic floras of Victoria. Parts I and II. *Geological Survey of Victoria Memoir* **28**, 1–310.

Gould RE 1980. The coal-forming flora of the Walloon coal measures. *Australian Coal Geology* **1**, 83–105.

Grant-Mackie JA, Aita Y, Balme BE, Campbell HJ, Challinor AB, MacFarlan DAB, Molnar RE, Stevens GR and Thulborn RA 2000. Jurassic palaeobiogeography of Australasia. *Memoir of the Association of Australasian Palaeontologists* **23**, 311–353.

McLoughlin S and Drinnan AN 1995. A Middle Jurassic flora from the Walloon Coal Measures, Mutdapilly, Queensland, Australia. *Memoirs of the Queensland Museum* **38**, 257–272.

McLoughlin S and Pott C 2009. The Jurassic flora of Western Australia. *GFF* **131**, 113–136.

Tidwell WD, Kim J-M and Kimura T 1987. Mid-Mesozoic leaves from near Ida Bay, southern Tasmania, Australia. *Papers and Proceedings of the Royal Society of Tasmania* **121**, 159–169.

Turner S, Bean LB, Dettmann M, McKellar JL, McLoughlin S and Thulborn T 2009. Australian Jurassic sedimentary and fossil successions: current work and future prospects for marine and non-marine correlation. *GFF* **131**, 49–70.

White ME 1981. Revision of the Talbragar fish bed flora (Jurassic) of New South Wales. *Records of the Australian Museum* **33**, 695–721.

Jurassic marine and non-marine invertebrates

Brunnschweiller RO 1960. Marine fossils from the Upper Jurassic and Lower Cretaceous of Dampier Peninsula, Western Australia. *Bureau of Mineral Resources, Geology and Geophysics Bulletin* **59**, 1–48.

Coleman PJ and Skwarko SK 1967. Lower Triassic and Middle Jurassic fossils at Enanty Hill, Mingenew, Perth Basin, Western Australia. *Bureau of Mineral Resources, Geology and Geophysics Bulletin* **92**, 197–214.

Grant-Mackie JA, Aita Y, Balme BE, Campbell HJ, Challinor AB, MacFarlan DAB, Molnar RE, Stevens GR and Thulborn RA 2000. Jurassic palaeobiogeography of Australasia. *Memoir of the*

Association of Australasian Palaeontologists **23**, 311–353.

Hall RL 1989. Lower Bajocian ammonites (Middle Jurassic, Soninniidae) from the Newmarracarra Limestone, Western Australia. *Alcheringa* **13**, 1–20.

Hocknull SA 2000. Mesozoic freshwater and estuarine bivalves from Australia. *Memoirs of the Queensland Museum* **45**, 405–426.

Jell PA 2004. Fossil insects of Australia. *Memoirs of the Queensland Museum* **50**, 1–123.

Skwarko SK 1974. Jurassic fossils of Western Australia. I: Bajocian Bivalvia of the Newmarracarra Limestone and the Kojareena Sandstone. *Bureau of Mineral Resources, Geology and Geophysics Bulletin* **150**, 57–110.

Jurassic non-marine fish

Bean LB 2006. The leptolepid fish *Cavenderichthys talbragarensis* (Woodward, 1895) from the Talbragar Fish Bed (Late Jurassic) near Gulgong, New South Wales. *Records of the Western Australian Museum* **23**, 43–76.

Dulhunty JA and Eadie J 1969. Geology of the Talbragar Fossil Fish Bed Area. *Journal and Proceedings of the Royal Society of New South Wales* **102**, 1–4.

Long JA 1991. The long history of Australian fossil fishes. In *Vertebrate Palaeontology of Australasia.* (Eds Vickers-Rich P, Monaghan JM, Baird RF and Rich TH) pp. 337–428. Pioneer Design Studio, Monash University, Melbourne.

López-Arbarello A 2004. The record of Mesozoic fishes from Gondwana (excluding India and Madagascar). In *Mesozoic Fishes 3: Systematics, Paleoenvironments and Biodiversity.* (Eds Arratia G and Tintori A) pp. 597–624. Verlag Dr Friedrich Pfeil, München.

Thies D and Turner S 2002. Jurassic actinopterygian fish from Monto, south-east Queensland. *Alcheringa* **25**, 381–386.

Turner S and Rozefelds A 1987. First Jurassic actinopterygian fish from Queensland. *Memoirs of the Queensland Museum* **25**, 201–206.

Turner S, Bean LB, Dettmann M, McKellar JL, McLoughlin S and Thulborn T 2009. Australian Jurassic sedimentary and fossil successions: current work and future prospects for marine and non-marine correlation. *GFF* **131**, 49–70.

Wade RT 1941. Australian Jurassic fishes of New South Wales. *Journal and Proceedings of the Royal Society of New South Wales* **75**, 71–84.

Woodward AS 1895. The fossil fishes of the Talbragar Beds (Jurassic?). *Memoirs of the Geological Survey of New South Wales, Palaeontology* **9**, 1–31.

Jurassic amphibians

Longman HA 1941. A Queensland fossil amphibian (*Austropelor*). *Memoirs of the Queensland Museum* **12**, 29–32.

Warren AA and Hutchinson MN 1983. The last labyrinthodont? A new brachyopoid (Amphibia, Temnospondyli) from the Early Jurassic Evergreen Formation of Queensland, Australia. *Philosophical Transactions of the Royal Society of London B* **303**, 1–62.

Jurassic plesiosaurs

Bartholomai AL 1966. The discovery of plesiosaurian remains in freshwater sediments in Queensland. *Australian Journal of Science* **28**, 437–438.

Kear BP 2004. Biogeographic and biostratigraphic implications of Australian Mesozoic marine reptiles. *Australian Biologist* **17**, 4–22.

Long JA and Cruickshank ARI 1998. Further records of plesiosaurian reptiles of Jurassic and Cretaceous age from Western Australia. *Records of the Western Australian Museum* **19**, 47–55.

Thulborn RH and Warren A 1980. Early Jurassic plesiosaurs from Australia. *Nature* **285**, 224–225.

Jurassic dinosaurs

Bartholomai A 1966. Fossil footprints in Queensland. *Australian Natural History* **15**, 147–150.

Long JA 1992. First dinosaur bones from Western Australia. *The Beagle* **9**, 21–27.

Long JA 1998. *Dinosaurs of Australia and New Zealand and Other Animals of the Mesozoic Era.* University of New South Wales Press, Sydney.

Long JA and Molnar RE 1998. A new Jurassic theropod dinosaur from Western Australia. *Records of the Western Australian Museum* **19**, 21–129.

Longman HA 1927. The giant dinosaur *Rhoetosaurus brownei. Memoirs of the Queensland Museum* **9**, 1–18.

Longman HA 1926. A giant dinosaur from Durham Downs, Queensland. *Memoirs of the Queensland Museum* **8**, 183–194.

Molnar RE 1991. Fossil reptiles in Australia. In *Vertebrate Palaeontology of Australasia.* (Eds Vickers-Rich P, Monaghan JM, Baird RF and Rich TH) pp. 605–702. Pioneer Design Studio, Monash University, Melbourne.

Thulborn RA 1994. Ornithopod dinosaur tracks from the Lower Jurassic of Queensland. *Alcheringa* **18**, 247–258.

Significant Australian Lower Cretaceous marine deposits and localities

Alexander EM and Sansome A 1996. Lithostratigraphy and environments of deposition. In *The Petroleum Geology of South Australia. Vol. 2: Eromanga Basin, South Australia.* (Eds Alexander EM and Hibbert JE) pp. 49–86. Department of Mines and Energy Report Book 96/20.

Burger D 1986. Palynology, cyclic sedimentation and palaeoenvironments in the Late Mesozoic of the Eromanga Basin. *Geological Society of Australia, Special Publication* **12**, 53–70.

Burger D 1988. Early Cretaceous environments in the Eromanga Basin: palynological evidence from GSQ Wyandra-1 corehole. *Memoirs of the Association of Australasian Palaeontologists* **5**, 173–186.

Burton GR and Mason AJ 1998. Controls on opal localisation in the White Cliffs area. *Quarterly Notes, Geological Survey of New South Wales* **107**, 1–11.

De Lurio JL and Frakes LA 1999. Glendonites as a palaeoenvironmental tool: implications for Early Cretaceous high-latitude climates in Australia. *Geochimica et Cosmochimica Acta* **63**, 1039–1048.

Embleton BJJ 1984. Australia's global setting: past global settings. In *Phanerozoic Earth History of Australia.* (Ed. Veevers JJ) pp. 11–17. Clarendon Press, Oxford.

Frakes LA and Francis JE 1988. A guide to Phanerozoic cold polar climates from high-latitude ice-rafting in the Cretaceous. *Nature* **333**, 547–549.

Frakes LA and Francis JE 1990. Cretaceous palaeoclimates. In *Cretaceous Resources, Events and Rhythms.* (Eds Ginsberg RN and Beaudoin B) pp. 273–287. Kluwer Academic Publishers, Dordrecht.

Frakes LA, Francis JE, Alley NF and Deynoux M 1995. Early Cretaceous ice rafting and climate zonation in Australia. *International Geology Review* **37**, 567–583.

Hocking RM, Moors HT and Van de Graff WJE 1987. Geology of the Carnarvon Basin. *Geological Survey of Western Australia, Bulletin* **133**, 1–289.

Idnurm M 1985. Late Mesozoic and Cenozoic palaeomagnetism of Australia. I. A redetermined apparent polar wander path. *Geophysical Journal of the Royal Astronomical Society* **83**, 399–418.

Johns RK 1968. Geology and mineral resources of the Andamooka-Torrens area. *Geological Survey of South Australia, Bulletin* **41**, 7–103.

Krieg GW and Rogers PA 1995. Stratigraphy: marine succession. In *The Geology of South Australia. Vol. 2. The Phanerozoic.* (Eds Drexel JF and Preiss WV) pp. 112–123. Geological Survey of South Australia, Bulletin 54.

McLoughlin S, Haig DW, Backhouse J, Holmes MA, Ellis G, Long JA and McNamara KJ 1995. Oldest Cretaceous sequence, Giralia Anticline, Carnarvon Basin, Western Australia: late Hauterivian-Barremian. *Journal of Australian Geology and Geophysics* **15**, 445–468.

Moore PS and Pitt GM 1985. Cretaceous subsurface stratigraphy of the south-western Eromanga Basin: a review. *South Australian Department of Mines and Energy, Special Publication* **5**, 269–286.

Moore PS, Pitt GM and Dettmann ME 1986. The Early Cretaceous Coorikiana Sandstone and Toolebuc Formation: their recognition and stratigraphic relationship in the south-western Eromanga Basin. In *Contributions to the Geology and Hydrocarbon Potential of the Eromanga Basin.* (Eds Gravestock DI, Moore PS and Pitt GM) pp. 97–114. *Geological Society of Australia, Special Publication* **12**.

Pirrie D, Doyle P, Marshall JD and Ellis G 1995. Cool Cretaceous climates: new data from the Albian of Western Australia. *Journal of the Geological Society, London* **152**, 739–742.

Price GD, Valdes PJ and Sellwood BW 1998. A comparison of GCM simulated Cretaceous 'greenhouse' and 'icehouse' climates: implications for the sedimentary record. *Palaeogeography, Palaeoclimatology, Palaeoecology* **142**, 123–138.

Robertson RS and Scott DC 1990. Geology of the Coober Pedy precious stones field. Results of investigations, 1981–1986. *Geological Survey of South Australia, Report of Investigations* **56**, 1–55.

Schneider SH, Thompson SL and Barron EJ 1985. Mid-Cretaceous continental surface temperatures: are high CO_2 concentrations needed to stimulate above-freezing winter conditions? In *The Carbon Cycle and Atmospheric CO_2: Natural Variations Archaean to Present.* (Eds Dundquist ET and Broecher WS) pp. 554–559. *American Geophysical Union, Geophysical Monograph* **32**.

Selwood BW, Price GD and Valdes PJ 1994. Cooler estimates of Cretaceous temperatures. *Nature* **370**, 453–455.

Sheard MJ 1990. Glendonites from the southern Eromanga Basin in South Australia: palaeoclimatic indicators for Cretaceous ice. *Geological Survey of South Australia, Quarterly Geological Notes* **114**, 17–23.

Wopfner H, Freytag IB and Heath GR 1970. Basal Jurassic-Cretaceous rocks of the western Great Artesian Basin, South Australia: stratigraphy and environment. *Association of Australian Petroleum Geologists Bulletin* **54**, 383–416.

Early Cretaceous marine invertebrates

Day RW 1969. The Early Cretaceous of the Great Artesian Basin. In *Stratigraphy and Palaeontology Essays in Honour of Dorothy Hill.* (Ed. Campbell KSW) pp. 140–173. Australian National University Press, Canberra.

Etheridge R Jr 1902. A monograph on the Cretaceous Mollusca of South Australia and the Northern Territory. *Memoirs of the Royal Society of South Australia* **2**, 1–54.

Etheridge R Jr 1902. A monograph on the Cretaceous invertebrate fauna of New South Wales. *Memoirs of the Geological Survey of New South Wales* **11**, 1–50.

Etheridge R Jr 1907. Lower Cretaceous fossils from the sources of the Barcoo, Ward and Nile Rivers, south central Queensland. Pt I: Annelida, Pelecypoda and Gasteropoda. *Records of the Australian Museum* **6**, 317–329.

Etheridge R Jr 1907. Lower Cretaceous fossils from the sources of the Barcoo, Ward and Nile Rivers, south central Queensland. Pt II: Cephalopoda. *Records of the Australian Museum* **7**, 235–240.

Etheridge R Jr 1920. Small Gasteropoda from the Lower Cretaceous of Queensland. *Publications of the Geological Survey of Queensland* **269**, 8–21.

Glaessner MF 1980. New Cretaceous and Tertiary crabs (Crustacea: Brachyura) from Australia and New Zeland. *Transactions of the Royal Society of South Australia* **104**, 171–192.

Henderson RA 1990. Late Albian ammonites from the Northern Territory. *Alcheringa* **14**, 109–148.

Henderson RA, Crampton JS, Dettmann ME, Douglas JG, Haig D, Shafik S, Stilwell JD and Thulborn RA 2000. Biogeographical observations on the Cretaceous biota of Australasia. *Memoirs of the Australasian Association of Palaeontologists* **23**, 355–404.

Ludbrook NH 1966. Cretaceous biostratigraphy of the Great Artesian Basin in South Australia.

Geological Survey of South Australia, Bulletin **40**, 1–223.

McNamara KJ 1978. *Myloceras* (Ammonoidea) from the Albian of central Queensland. *Alcheringa* **2**, 231–242.

McNamara KJ 1980. Heteromorph ammonites from the Albian of South Australia. *Transactions of the Royal Society of South Australia* **104**, 145–159.

McNamara KJ 1985. A new micromorph ammonite genus from the Albian of South Australia. *Special Publication of the South Australian Department of Mines and Energy* **5**, 263–268.

Moore C 1870. Australian Mesozoic geology and palaeontology. *Quarterly Journal of the Geological Society of London* **26**, 226–261.

Skwarko SK 1967. Lower Cretaceous Mollusca of the Great Artesian Basin type in the Gibson Desert, central Western Australia. *Bulletin of the Bureau of Mineral Resources, Geology and Geophysics, Australia* **75**, 1–35.

Wade M 1993. New Kelaenida and Vampyromorpha: Cretaceous squid from Queensland. *Memoirs of the Association of Australasian Palaeontologists* **15**, 353–374.

Whitehouse FW 1926. The Cretaceous Ammonoidea of eastern Australia. *Memoirs of the Queensland Museum* **8**, 195–242.

Early Cretaceous marine sharks and bony fish

Bartholomai A 1969. The Lower Cretaceous elopid fish *Pachyrhizodus marathonensis* (Etheridge Jnr). In *Stratigraphy and Palaeontology Essays in Honour of Dorothy Hill.* (Ed. Campbell KSW) pp. 249–263. Australian National University Press, Canberra.

Bartholomai A 2004. The large aspidorhynchid fish *Richmondichthys sweeti* (Etheridge Jr and Smith Woodward, 1891) from Albian marine deposits of Queensland, Australia. *Memoirs of the Queensland Museum* **49**, 521–536.

Cappetta H 1987. *Chondrichtyes II: Mesozoic and Cenozoic Elasmobranchii. Handbook of Paleoichthyology Vol. 3B.* Gustav Fisher Verlag, Stuttgart.

Kear BP 2007. First record of a pachycormid fish (Actinopterygii: Pachycormiformes) from the Early Cretaceous of Australia. *Journal of Vertebrate Paleontology* **27**, 1033–1038.

Kemp N 1991. Chondrichthyans in the Cretaceous and Tertiary of Australia. In *Vertebrate Palaeontology of Australasia.* (Eds Vickers-Rich P, Monaghan JM, Baird RF and Rich TH) pp. 497–568. Pioneer Design Studio, Monash University, Melbourne.

Lees TA 1986. A new chimaeroid *Ptychoptychion tayyo* gen. et sp. nov. (Pisces: Holocephali) from the marine Cretaceous of Queensland. *Alcheringa* **10**, 187–193.

Lees TA 1990. A probable neoteleost *Dugaldia emmilta* gen. et sp. nov., from the Lower Cretaceous of Queensland, Australia. *Memoirs of the Queensland Museum* **28**, 79–88.

Lees TA and Bartholomai A 1987. Study of a Lower Cretaceous actinopterygian (Class Pisces) *Cooyoo australis* from Queensland, Australia. *Memoirs of the Queensland Museum* **8**, 16–28.

Long JA 1985. A new Cretaceous chimaerid (Pisces: Holocephali) from South Australia. *Transactions of the Royal Society of South Australia* **109**, 49–53.

Long JA 1991. The long history of Australian fossil fishes. In *Vertebrate Palaeontology of Australasia.* (Eds Vickers-Rich P, Monaghan JM, Baird RF and Rich TH) pp. 337–428. Pioneer Design Studio, Monash University, Melbourne.

Longman HA 1932. A new Cretaceous fish. *Memoirs of the Queensland Museum* **10**, 89–97.

Turner S and Rozefelds A 1994. Tip of the pyramid? Cretaceous megasharks from Australia. *Records of the South Australian Museum* **27**, 262–263.

Early Cretaceous sea turtles

Gaffney ES 1981. A review of the fossil turtles of Australia. *American Museum Novitates* **2720**, 1–38.

Kear BP 2003. Cretaceous marine reptiles of Australia: a review of taxonomy and distribution. *Cretaceous Research* **24**, 277–303.

Kear BP 2006. First gut contents in a Cretaceous sea turtle. *Biology Letters* **2**, 113–115.

Kear BP 2006. Reassessment of *Cratochelone berneyi* Longman, 1915, a giant Early Cretaceous sea turtle from Australia. *Journal of Vertebrate Paleontology* **26**, 779–783.

Kear BP and Lee MSY 2006. A primitive protostegid from Australia and early sea turtle evolution. *Biology Letters* **2**, 116–119.

Longman HA 1915. On a giant turtle from the Queensland Lower Cretaceous. *Memoirs of the Queensland Museum* **3**, 24–29.

Early Cretaceous ichthyosaurs

Choo B 1999. Cretaceous ichthyosaurs from Western Australia. *Records of the Western Australian Museum, Supplement* **57**, 207–218.

Kear BP 2003. Cretaceous marine reptiles of Australia: a review of taxonomy and distribution. *Cretaceous Research* **24**, 277–303.

Kear BP 2005. Cranial morphology of *Platypterygius longmani* Wade, 1990 (Reptilia: Ichthyosauria) from the Lower Cretaceous of Australia. *Zoological Journal of the Linnean Society* **145**, 583–622.

Kear BP, Boles WE and Smith ET 2003. Unusual gut contents in a Cretaceous ichthyosaur. *Proceedings of the Royal Society of London, Series B* **270**, S206–S208.

Maisch MW and Matzke AT 2000. *The Ichthyosauria.* Stuttgarter Beiträge zur Naturkunde, Serie B (Geologie und Paläontologie) **298**, 1–159.

McCoy F 1867. On the occurrence of *Ichthyosaurus* and *Plesiosaurus* in Australia. *Annual Magazine of Natural History* **19**, 355–356.

McGowan C 1972. The systematics of Cretaceous ichthyosaurs with particular reference to the material from North America. *Contributions to Geology, University of Wyoming* **11**, 9–29.

McGowan C and Motani R 2003. *Ichthyopterygia. Handbuch der Paläoherpetologie Part 8.* Verlag Dr Friedrich Pfeil, München.

Motani R 1999. Phylogeny of the Ichthyopterygia. *Journal of Vertebrate Paleontology* **19**, 473–496.

Wade M 1984. *Platypterygius australis*, an Australian Cretaceous ichthyosaur. *Lethaia* **17**, 99–113.

Wade M 1990. A review of the Australian Cretaceous longipinnate ichthyosaur *Platypterygius*, (Ichthyosauria: Ichthyopterygia). *Memoirs of the Queensland Museum* **28**, 115–137.

Early Cretaceous plesiosaurs

Cruickshank ARI and Long JA 1997. A new species of pliosaurid reptile from the Early Cretaceous Birdrong Sandstone of Western Australia. *Records of the Western Australian Museum* **18**, 263–276.

Cruickshank ARI, Fordyce RE and Long JA 1999. Recent developments in Australasian sauropterygian palaeontology (Reptilia, Sauropterygia). *Records of the Western Australian Museum, Supplement* **57**, 201–205.

Etheridge R 1897. An Australian sauropterygian (*Cimoliasaurus*), converted into precious opal. *Records of the Australian Museum* **3**, 21–29.

Etheridge R 1904. A second sauropterygian converted into opal from the Upper Cretaceous of White Cliffs, New South Wales. With indications of ichthyopterygians at the same locality. *Records of the Australian Museum* **5**, 306–316.

Freytag IB 1964. Reptilian vertebral remnants from Lower Cretaceous strata near Oodnadatta. *Geological Survey of South Australia, Quarterly Geological Notes* **10**, 1–2.

Kear BP 2001. Elasmosaur (Reptilia: Plesiosauria) basicranial remains from the Early Cretaceous of Queensland. *Records of the South Australian Museum* **34**, 127–133.

Kear BP 2002. Reassessment of the Early Cretaceous plesiosaur *Cimoliasaurus maccoyi* Etheridge, 1904 (Reptilia: Sauropterygia) from White Cliffs, New South Wales. *Australian Journal of Zoology* **50**, 671–685.

Kear BP 2002. Darwin Formation (Early Cretaceous, Northern Territory) marine reptile remains in the South Australian Museum. *Records of the South Australian Museum* **35**, 33–47.

Kear BP 2005. A new elasmosaurid plesiosaur from the Lower Cretaceous of Queensland, Australia. *Journal of Vertebrate Paleontology* **25**, 792–805.

Kear BP 2005. Marine reptiles from the Lower Cretaceous (Aptian) deposits of White Cliffs, south-eastern Australia: implications of a high-latitude cold water assemblage. *Cretaceous Research* **26**, 769–782.

Kear BP 2006. Marine reptiles from the Lower Cretaceous of South Australia: elements of a high-latitude cold water assemblage. *Palaeontology* **49**, 837–856.

Kear BP 2006. Plesiosaur remains from Cretaceous high-latitude non-marine deposits in south-eastern Australia. *Journal of Vertebrate Paleontology* **26**, 196–199.

Kear BP 2007. A juvenile pliosauroid plesiosaur from the Lower Cretaceous of South Australia. *Journal of Paleontology* **81**, 154–162.

Kear BP 2007. Taxonomic clarification of the Australian elasmosaurid *Eromangasaurus*, with reference to other austral elasmosaur taxa. *Journal of Vertebrate Paleontology* **27**, 241–246.

Kear BP, Schroeder NI and Lee MSY 2006. An archaic crested plesiosaur in opal from the Lower Cretaceous high-latitude deposits of Australia. *Biology Letters* **2**, 615–619.

Kear BP, Schroeder NI, Vickers-Rich P and Rich TH 2006. Early Cretaceous high-latitude marine reptile assemblages from southern Australia. *Paludicola* **5**, 200–205.

Long JA and Cruickshank ARI 1998. Further records of plesiosaurian reptiles of Jurassic and Cretaceous age from Western Australia. *Records of the Western Australian Museum* **19**, 47–55.

Longman HA 1924. A new gigantic marine reptile from the Queensland Cretaceous, *Kronosaurus queenslandicus* new genus and species. *Memoirs of the Queensland Museum* **8**, 26–28.

McHenry C, Cook A and Wroe S 2005. Bottom-feeding plesiosaurs. *Science* **310**, 75.

Murray PF 1987. Plesiosaurs from Albian aged Darwin Formation siltstones near Darwin, Northern Territory, Australia. *The Beagle* **4**, 95–102.

Persson PO 1960. Early Cretaceous plesiosaurians (Reptilia) from Australia. *Lunds Universitets Årsskrift* **56**, 1–23.

Thulborn T and Turner S 1993. An elasmosaur bitten by a pliosaur. *Modern Geology* **18**, 489–501.

Welles SP 1962. A new species of elasmosaur from the Aptian of Colombia, and a review of the Cretaceous plesiosaurs. *University of California, Publications in Geological Sciences* **44**, 1–96.

White TE 1935. On the skull of *Kronosaurus queenslandicus* Longman. *Occasional papers of the Boston Society of Natural History* **8**, 219–228.

Significant Australian Lower Cretaceous non-marine deposits and localities

Alexander EM and Krieg GW 1995. Stratigraphy: lower non-marine succession. In *The Geology of South Australia. Vol. 2. The Phanerozoic.* (Eds Drexel JF and Preiss WV) pp. 105–112. *Geological Survey of South Australia, Bulletin* **54**.

Alley NF and Frakes LA 2003. First known Cretaceous glaciation: Livingston Tillite Member of the Cadna-owie Formation, South Australia. *Australian Journal of Earth Sciences* **50**, 139–144.

Alley NF and Pledge NS 2000. The plants, animals and environments of the last 280 (290) million years. In *Lake Eyre Monograph Series. Vol. 5.* (Ed. Slaytor WJH) pp. 35–82. Royal Geographic Society of South Australia, Adelaide.

Bryan SE, Constantine AE, Stephens CJ, Ewart A, Schon RW and Parianos J 1997. Early Cretaceous volcano-sedimentary successions along the eastern Australian continental margin: implications for the break-up of eastern Australia. *Earth and Planetary Science Letters* **153**, 85–102.

Burger D 1980. Palynology of the Lower Cretaceous Surat Basin. *Bureau of Mineral Resources Australia, Geology and Geophysics Bulletin* **189**, 1–106.

Byrnes J 1977. Notes on the Rolling Downs Group in the Milparinka, White Cliffs and Angledool 1:250 000 Sheet Areas. *Geological Survey of New South Wales, Report* GS **1977/005**, 1–17.

Constantine A, Chinsamy A, Rich TH and Vickers-Rich P 1998. Periglacial environments and polar dinosaurs. *South African Journal of Science* **94**, 137–141.

Exon NF 1976. Geology of the Surat Basin in Queensland. *Bureau of Mineral Resources Australia, Bulletin* **166**, 1–104.

Exon NF and Senior BR 1976. The Cretaceous geology of the Eromanga and Surat Basins in Queensland. *Bureau of Mineral Resources Australia, Journal of Australian Geology and Geophysics* **1**, 33–50.

Gleadow AJW and Duddy IR 1980. Early Cretaceous volcanism and the early breakup history of south-eastern Australia: evidence from fission-track dating of volcanoclastic sediments. In *Proceedings of the Fifth International Gondwana Symposium, Wellington, New Zealand.* (Eds Creswell MM and Vella P) pp. 295–300. Balkema, Rotterdam.

Gregory RT, Douthitt CB, Duddy IR, Rich PV and Rich TH 1989. Oxygen isotopic composition of carbonate concretions from the Lower Cretaceous of Victoria, Australia: implications for the evolution of meteoric waters on the Australian continent in a paleopolar environment. *Earth and Planetary Science Letters* **92**, 27–42.

Wagstaff BE and McEwen Mason J 1989. Palynological dating of Lower Cretaceous coastal vertebrate localities, Victoria, Australia. *National Geographic Society Research Reports* **5**, 54–63.

Whitelaw MJ 1993. Paleomagnetic paleolatitude determinations for the Cretaceous vertebrate localities of south-eastern Australia: high-latitude dinosaur faunas. *Journal of Vertebrate Paleontology* **3**, 62A.

Early Cretaceous land plants

Burger D 1990. Early Cretaceous angiosperms from Queensland, Australia. *Review of Palaeobotany and Palynology* **65**, 53–63.

Burger D 1993. Early and middle Cretaceous angiosperm pollen grains from Australia. *Review of Palaeobotany and Palynology* **78**, 183–234.

Cantrill DJ and Poole I 2002. Cretaceous patterns of floristic change in the Antarctic Peninsula. In *Palaeobiogeography and Biodiversity Change: The Ordovician and Mesozoic-Cenozoic Radiations.* (Eds Crame JA and Owen AW) pp. 141–152. *Geological Society Special Publication* **194**.

Cantrill DJ and Webb JA 1987. A reappraisal of *Phyllopteroides* Medwell (Osmundaceae) and its stratigraphic significance in the Lower Cretaceous of eastern Australia. *Alcheringa* **11**, 59–85.

Dettmann ME 1963. Upper Mesozoic microfloras from south-eastern Australia. *Proceedings of the Royal Society of Victoria* **77**, 1–148.

Dettmann ME 1986. Early Cretaceous palynoflora of subsurface strata correlative with the Koonwarra Fossil Bed, Victoria. *Memoirs of the Association of Australasian Palaeontologists* **3**: 79–110.

Dettmann ME and Jarzen DM 1990. The Antarctic/Australian Rift Valley: Late Cretaceous cradle of north-eastern Australian relicts? *Review of Palaeobotany and Palynology* **65**, 131–144.

Dettmann ME and Playford G 1969. Palynology of the Australian Cretaceous: a review. In *Stratigraphy and Palaeontology. Essays in Honour of Dorothy Hill.* (Ed. Campbell KSW) pp. 174–210. Australian National University Press, Canberra.

Dettmann ME, Molnar RE, Douglas JG, Burger D, Fielding C, Clifford HT, Francis J, Jell P, Rich T, Wade M, Rich PV, Pledge N, Kemp A and Rozefields A 1992. Australian Cretaceous terrestrial faunas and floras: biostratigraphic and biogeographic implications. *Cretaceous Research* **13**, 207–262.

Douglas JG and Williams GE 1982. Southern polar forests: the Early Cretaceous floras of Victoria and their palaeoclimatic significance. *Palaeogeography, Palaeoclimatology, Palaeoecology* **39**, 171–185.

Drinnan AN and Chambers TC 1986. Flora of the Lower Cretaceous Koonwarra Fossil Beds (Korumburra Group), South Gippsland, Victoria. *Memoirs of the Australasian Association of Palaeontologists* **3**, 1–77.

McLoughlin S 1996. Early Cretaceous macrofloras of Western Australia. *Records of the Western Australian Museum* **18**, 19–65.

McLoughlin S, Tosolini A-MP and Drinnan AN 2000. Revision of an Early Cretaceous macroflora from the Maryborough Formation, Maryborough Basin, Queensland, Australia. *Memoirs of the Queensland Museum* **45**, 483–503.

McLoughlin S, Tosolini A-MP, Nagalingum N and Drinnan AN 2002. The Early Cretaceous (Neocomian) flora and fauna of the lower Strzelecki Group, Gippsland Basin, Victoria, Australia. *Memoirs of the Association of Australasian Palaeontologists* **26**, 1–144.

Mutter JC, Hegarty KA, Cande SC and Wessel JK 1985. Break-up between Australia and Antarctica: a brief review in the light of new data. *Tectonophysics* **114**, 255–279.

Nagalingum NS, Drinnan AN, Lupia R and McLoughlin S 2002. Fern spore diversity and abundance in Australia during the Cretaceous. *Review of Palaeobotany and Palynology* **119**, 69–92.

Early Cretaceous non-marine invertebrates

Hamilton-Bruce RJ and Kear BP 2006. A new fossil non-marine snail (Gastropoda) from the Lower Cretaceous (Albian, Griman Creek Formation) of eastern Australia. *Molluscan Research* **26**, 84–88.

Hamilton-Bruce RJ, Kear BP and Smith BJ 2004. A new species of non-marine Early Cretaceous gastropod from Lightning Ridge, New South Wales. *Alcheringa* **28**, 485–492.

Hamilton-Bruce RJ, Smith BJ and Gowlett-Holmes KL 2002. Descriptions of a new genus and two new species of viviparid snails (Mollusca: Gastropoda: Viviparidae) from the Early Cretaceous (middle-Late Albian) Griman Creek Formation of Lightning Ridge, northern New South Wales. *Records of the South Australian Museum* **35**, 193–203.

Hocknull SA 1997. Cretaceous freshwater bivalves from Queensland. *Memoirs of the Queensland Museum* **42**, 223–226.

Hocknull SA 2000. Mesozoic freshwater and estuarine bivalves from Australia. *Memoirs of the Queensland Museum* **45**, 405–426.

Jell PA 2004. Fossil insects of Australia. *Memoirs of the Queensland Museum* **50**, 1–123.

Jell PA and Duncan PM 1986. Invertebrates, mainly insects, from the freshwater, Lower Cretaceous, Koonwarra Fossil Bed (Korumburra Group), South Gippsland, Victoria. *Memoirs of the Association of Australasian Palaeontologists* **3**, 111–205.

Kear BP, Hamilton-Bruce RJ, Smith BJ and Gowlett-Holmes KL 2003. A reassessment of Australia's oldest freshwater snail *Viviparus* (?) *albascopularis* Etheridge, 1902 (Mollusca: Gastropoda: Viviparidae), from the Lower Cretaceous (Aptian, Wallumbilla Formation) of White Cliffs, New South Wales. *Molluscan Research* **23**, 149–158.

McMichael DF 1957. A review of fossil freshwater mussels (Pelecypoda: Unionidae) of Australasia. *Proceedings of the Linnean Society of New South Wales* **81**, 222–242.

Smith E 1999. *Black Opal Fossils of Lightning Ridge.* Kangaroo Press, Sydney.

Early Cretaceous non-marine sharks and fish

Hall TS 1900. A new genus and a new species of fish from the Mesozoic rocks of Victoria. *Proceedings of the Royal Society of Victoria* **12**, 147–151.

Kemp A 1993. *Ceratodus diutinus*, a new ceratodont from Cretaceous and late Oligocene-medial Miocene deposits in Australia. *Journal of Paleontology* **67**, 883–888.

Kemp A 1997. A revision of Australian Mesozoic and Cenozoic lungfish of the family Neoceratodontidae (Osteichthyes: Dipnoi), with a description of four new species. *Journal of Paleontology* **71**, 713–733.

Kemp A 1997. Four species of *Metaceratodus* (Osteichthyes: Dipnoi, Family Ceratodontidae) from Australian Mesozoic and Cenozoic deposits. *Journal of Vertebrate Paleontology* **17**, 26–33.

Kemp A and Molnar RE 1981. *Neoceratodus forsteri* from the Lower Cretaceous of New South Wales, Australia. *Journal of Paleontology* **55**, 211–217.

Long JA 1991. The long history of Australian fossil fishes. In *Vertebrate Palaeontology of Australasia.* (Eds Vickers-Rich P, Monaghan JM, Baird RF and Rich TH) pp. 337–428. Pioneer Design Studio, Monash University, Melbourne.

Smith E and Smith R 1999. *Black Opal Fossils of Lightning Ridge.* Kangaroo Press, Sydney.

Waldman M 1971. Fish from the freshwater Lower Cretaceous of Victoria, Australia, with comments on the palaeoenvironment. *Special Papers in Paleontology* **9**, 1–124.

Waldman M 1971. A reexamination of *Psilichthys selwyni* Hall, from the Lower Cretaceous of Victoria. *Proceedings of the Royal Society of Victoria* **84**, 263–266.

Early Cretaceous amphibians

Warren AA, Kool L, Cleeland M, Rich TH and Vickers-Rich P 1997. The last labyrinthodonts? *Palaeontographica* **247A**, 1–24.

Warren AA, Kool L, Cleeland M, Rich TH and Vickers-Rich P 1991. An Early Cretaceous labyrinthodont. *Alcheringa* **15**, 327–332.

Early Cretaceous non-marine turtles

Gaffney ES, Kool L, Brinkman DB, Rich TH and Vickers-Rich P 1998. *Otwayemys*, a new cryptodiran turtle from the Early Cretaceous of Australia. *American Museum Novitates* **3233**, 1–28.

Lapparent de Broin F de and Molnar RE 2001. Eocene chelid turtles from Redbank Plains, south-east Queensland. *Geodiversitas* **23**, 41–79.

Smith ET 2009. Terrestrial and freshwater turtles of Early Cretaceous Australia. Unpublished PhD thesis, University of New South Wales, Sydney.

Warren JW 1969. A fossil chelonian of probable Lower Cretaceous age from Victoria, Australia. *Memoirs of the National Museum of Victoria* **29**, 23–28.

Early Cretaceous lizards

Molnar RE 1991. Fossil reptiles in Australia. In *Vertebrate Palaeontology of Australasia.* (Eds Vickers-Rich P, Monaghan JM, Baird RF and Rich TH) pp. 605–702. Pioneer Design Studio, Monash University, Melbourne.

Rich TH and Rich PV 1989. Polar dinosaurs and biotas of the Early Cretaceous of south-eastern Australia. *National Geographic Research* **5**, 15–53.

Early Cretaceous crocodiles

Molnar RE 1980. Procoelous crocodile from the Lower Cretaceous of Lightning Ridge, NSW. *Memoirs of the Queensland Museum* **20**, 65–75.

Molnar RE and Willis PMA 2001. New crocodyliform material from the Early Cretaceous Griman Creek Formation, at Lightning Ridge, New South Wales. In *Crocodilian Biology and Evolution.* (Eds Grigg GC, Seebacher F and Franklin CE) pp. 75–82. Surrey Beatty and Sons, Sydney.

Early Cretaceous dinosaurs

Bartholomai A and Molnar RE 1981. *Muttaburrasaurus langdoni*, a new genus and species of iguanodontid (Ornithischia: Ornithopoda) dinosaur from the Lower Cretaceous of Queensland, Australia. *Memoirs of the Queensland Museum* **20**, 319–350.

Coombs WP Jr and Molnar RE 1981. Sauropods (Reptilia: Saurischia) from the Cretaceous of Queensland. *Memoirs of the Queensland Museum* **20**, 351–373.

Currie PJ, Vickers-Rich P and Rich TH 1996. Possible oviraptorosaur (Theropoda: Dinosauria) specimens from the Early Cretaceous Otway Group of Dinosaur Cove, Australia. *Alcheringa* **20**, 73–79.

Long JA 1998. *Dinosaurs of Australia and New Zealand and Other Animals of the Mesozoic Era.* University of New South Wales Press, Sydney.

Longman HA 1933. A new dinosaur from the Queensland Cretaceous. *Memoirs of the Queensland Museum* **13**, 133–144.

Molnar RE 1980. Australian late Mesozoic terrestrial tetrapods: some implications. *Memoirs de la Société Géologique de France* **139**, 131–143.

Molnar RE 1980. An ankylosaur (Ornithischia: Reptilia) from the Lower Cretaceous of southern Queensland. *Memoirs of the Queensland Museum* **20**, 77–87.

Molnar RE 1996. Observations on the Australian ornithopod dinosaur *Muttaburrasaurus. Memoirs of the Queensland Museum* **39**, 639–652.

Molnar RE 1996. Preliminary report on a new ankylosaur from the Early Cretaceous of Australia. *Memoirs of the Queensland Museum* **39**, 653–668.

Molnar RE 2001. Armour of the small ankylosaur *Minmi*. In *The Armoured Dinosaurs.* (Ed. Carpenter K) pp. 341–362. Indiana University Press, Bloomington.

Molnar RE 2001. An ankylosaurian cololite from the Lower Cretaceous of Queensland, Australia. In *The Armoured Dinosaurs.* (Ed. Carpenter K) pp. 399–412. Indiana University Press, Bloomington.

Molnar RE and Clifford HT 2000. Gut contents of a small ankylosaur. *Journal of Vertebrate Paleontology* **20**, 194–196.

Molnar RE, Flannery TF and Rich TH 1981. An allosaurid theropod dinosaur from the Early Cretaceous of Victoria, Australia. *Alcheringa* **5**, 141–146.

Molnar RE and Galton PM 1986. Hypsilophodontid dinosaurs from Lightning Ridge, New South Wales. *Geobios* **19**, 231–239.

Molnar RE and Frey E 1987. The paravertebral elements of the Australian ankylosaur *Minmi* (Reptilia: Ornithischia, Cretaceous). *Neues Jahrbuch für Geologie und Paläontologie Abhandlungen* **175**, 19–37.

Molnar RE and Pledge NS 1980. A new theropod dinosaur from South Australia. *Alcheringa* **4**, 281–287.

Molnar RE and Salisbury SW 2005. Observations on Cretaceous sauropods from Australia. In *Thunder-Lizards: The Sauropodomorph Dinosaurs.* (Eds Tidwell V and Carpenter K) pp. 454–465. Indiana University Press, Bloomington.

Rich TH 1996. Significance of polar dinosaurs in Gondwana. *Memoirs of the Queensland Museum* **39**, 711–717.

Rich TH and Rich PV 1989. Polar dinosaurs and biotas of the Early Cretaceous of south-eastern Australia. *National Geographic Society, Research Reports* **5**, 15–53.

Rich TH and Vickers-Rich P 1994. Neoceratopsians and ornithomimosaurs: dinosaurs of Gondwanan origin? *National Geographic Research and Exploration* **5**, 129–131.

Rich TH and Vickers-Rich P 1999. The hypsilophodontidae from south-eastern Australia. In *Proceedings of the Second Gondwanan Dinosaur Symposium. National Science Museum Monographs* **15**. (Eds Tomida Y, Rich TH and Vickers-Rich P) pp. 167–180. Tokyo.

Rich TH and Vickers-Rich P 2000. *Dinosaurs of Darkness.* University of Indiana Press, Bloomington.

Rich TH and Vickers-Rich P 2003. Protoceratopsian? ulnae from Australia. *Records of the Queen Victoria Museum* **113**, 1–12.

Rich PV, Rich TH, Wagstaff BE, McEwen Mason J, Douthitt CB, Gregory RT and Mason EA 1988. Evidence for low temperatures and biologic diversity in Cretaceous high latitudes of Australia. *Science* **242**, 1403–1406.

Salisbury S, Agnolin F, Ezcurra M and Pais D 2007. A critical reassessment of the Cretaceous non-avian dinosaur faunas of Australia and New Zealand. *Journal of Vertebrate Paleontology* **27** (Supplement 3), 138A.

Scanlon JD 2006. Dinosaurs and other Mesozoic reptiles of Australasia. In *Evolution and Biogeograpghy of Australasian Vertebrates.* (Eds Merrick JR, Archer M, Hickey GM and Lee MSY) pp. 265–290. Australian Science Publishing, Sydney.

Thulborn RA, Hamley T and Foulkes P 1994. Preliminary report on sauropod dinosaur tracks in the Broome Sandstone (Lower Cretaceous) of Western Australia. *Gaia* **10**, 85–96.

Vickers-Rich P 1996. Early Cretaceous polar tetrapods from the Great Southern Rift Valley, south-eastern Australia. *Memoirs of the Queensland Museum* **39**, 719–723.

Early Cretaceous pterosaurs

Hocknull SA and Cook AG 2007. Flights of fancy: on the trail of the elusive Aussie pterosaurs. *Australian Age of Dinosaurs* **5**, 26–35.

Molnar RE 1987. A pterosaur pelvis from western Queensland. *Alcheringa* **11**, 87–94.

Molnar RE 1991. Fossil reptiles in Australia. In *Vertebrate Palaeontology of Australasia.* (Eds Vickers-Rich P, Monaghan JM, Baird RF and Rich TH) pp. 605–702. Pioneer Design Studio, Monash University, Melbourne.

Molnar RE and Thulborn RA 1980. First pterosaur from Australia. *Nature* **288**, 361–363.

Molnar RE and Thulborn RA 2007. An incomplete pterosaur skull from the Cretaceous of north-central Queensland, Australia. *Arquivos do Museu Nacional* **65**, 461–470.

Rich TH and Vickers-Rich P 2000. *Dinosaurs of Darkness.* University of Indiana Press, Bloomington.

Early Cretaceous birds

Close RA, Vickers-Rich P, Trusler P, Chiappe LM, O'Connor J, Rich TH, Kool L and Komarower P 2009. Earliest Gondwanan bird from the Cretaceous of south-eastern Australia. *Journal of Vertebrate Paleontology* **29**, 616–619.

Molnar RE 1986. An enantiornithine bird from the Lower Cretaceous of Queensland. *Nature* **322**, 736–738.

Molnar RE 1999. Avian tibiotarsi from the Early Cretaceous of Lightning Ridge, New South Wales. In *Proceedings of the Second Gondwanan Dinosaur Symposium. National Science Museum Monographs* **15**. (Eds Tomida Y, Rich TH and Vickers-Rich P) pp. 197–209. Tokyo.

Talent JA, Duncan PM and Handley PL 1966. Early Cretaceous feathers from Victoria. *Emu* **66**, 81–86.

Early Cretaceous relict therapsids

Thulborn T and Turner S 2003. The last dicynodont: an Australia Cretaceous relict. *Proceedings of the Royal Society of London, Series B* **270**, 985–993.

Early Cretaceous mammals

Archer M, Flannery TF, Ritchie A and Molnar RE 1985. First Mesozoic mammal from Australia: an Early Cretaceous monotreme. *Nature* **318**, 363–366.

Clemens WA, Wilson GP and Molnar RE 2003. An enigmatic (Synapsid?) tooth from the Early Cretaceous of New South Wales, Australia. *Journal of Vertebrate Paleontology* **23**, 232–237.

Flannery TF, Archer M, Rich TH and Jones J 1995. A new family of monotremes from the Cretaceous of Australia. *Nature* **377**, 418–420.

Pridmore PA, Rich TH, Vickers-Rich P and Gambaryan PP 2005. A tachyglossid-like humerus from the Early Cretaceous of south-eastern Australia. *Journal of Mammalian Evolution* **12**, 359–378.

Rich TH, Flannery TF and Archer M 1989. A second Cretaceous mammalian specimen from Lightning Ridge, New South Wales, Australia. *Alcheringa* **13**, 85–88.

Rich TH, Flannery TF, Trusler P, Kool L, van Klaveren N and Vickers-Rich P 2001. A second tribophenic mammal from the Mesozoic of Australia. *Records of the Queen Victoria Museum* **110**, 1–9.

Rich TH, Vickers-Rich P, Constantine A, Flannery TF, Kool L and van Klaveren N 1997. A tribophenic mammal from the Mesozoic of Australia. *Science* **278**, 1438–1142.

Rich TH, Vickers-Rich P, Constantine A, Flannery TF, Kool L and van Klaveren N 1999. Early Cretaceous mammals from Flat Rocks, Victoria, Australia. *Records of the Queen Victoria Museum* **106**, 1–35.

Rich TH, Vickers-Rich P, Flannery TF, Kear BP, Cantrill D, Komarower P, Kool L, Pickering D, Trusler P, van Klaveren N and Fitzgerald EMG 2009. An Australian multituberculate and its palaeobiogeographical implications. *Acta Palaeontologica Polonica* **54**, 1–6.

Rich TH, Vickers-Rich P, Trusler P, Flannery TF, Cifelli R, Constantine A, Kool L and van Klaveren N 2001. Monotreme nature of the Australian Early Cretaceous mammal *Teinolophos*. *Acta Palaeontologica Polonica* **46**, 113–118.

Significant Australian Upper Cretaceous deposits and localities

Apthorpe MC 1979. Depositional history of the Upper Cretaceous of the Northwest Shelf, based on foraminifera. *APEA Journal* **19**, 74–89.

Belford DJ 1958. Stratigraphy and micropalaeontology of the Cretaceous of Western Australia. *Geologische Rundschau* **47**, 629–647.

Dentith MC, Bevan AWR, Backhouse J, Featherstone WE and Koeberl C 1999. Yallalie: a buried structure of possible impact origin in the Perth Basin, Western Australia. *Geological Magazine* **136**, 619–632.

Henderson RA 1998. Palaeoenvironmental and eustatic record of the mid-Cretaceous Bathurst Island Group, Money Shoals Platform, Northern Territory. *Palaeogeography, Palaeoclimatology, Palaeoecology* **138**, 115–138.

Hocking RM, Moors HT and Van de Graff WJE 1987. Geology of the Carnarvon Basin. *Geological Survey of Western Australia, Bulletin* **133**, 1–289.

McWae JRH, Playford PE, Lindner AW, Glenister BF and Balme BE 1958. The stratigraphy of Western Australia. *Journal of the Geological Society of Australia* **4**, 1–161.

Mory AJ, Haig DW, McLoughlin S and Hocking RM 2005. Geology of the northern Perth Basin, Western Australia: a field guide. *Western Australian Geological Survey, Record* **2005/9**, 1–71.

Mutter JC, Hegarty KA, Cande SC and Wessel JK 1985. Break-up between Australia and Antarctica: a brief review in the light of new data. *Tectonophysics* **114**, 255–279.

Playford PE, Cockbain AE and Low GH 1976. Geology of the Perth Basin, Western Australia. *Western Australia Geological Survey, Bulletin* **124**, 1–311.

Schafik S 1990. Late Cretaceous nannofossil biostratigraphy and biogeography of the Australian western margin. *Bureau of Mineral Resources, Geology and Geophysics, Report* **295**, 1–164.

Late Cretaceous land plants

Dettmann ME and Clifford HT 2000. Gemmae of the marchantiales from the Winton Formation (mid-Cretaceous), Eromanga Basin, Queensland. *Memoirs of the Queensland Museum* **45**, 285–292.

Dettmann ME, Clifford HT and Peters M 2009. *Lovellea wintonensis* gen. et sp. nov. – early Cretaceous (late Albian), anatomically preserved, angiosperm flowers and fruits from the Winton Formation, western Queensland, Australia. *Cretaceous Research* **30**, 339–335.

Dettmann ME, Molnar RE, Douglas JG, Burger D, Fielding C, Clifford HT, Francis J, Jell P, Rich T, Wade M, Rich PV, Pledge N, Kemp A and Rozefields A 1992. Australian Cretaceous terrestrial faunas and floras: biostratigraphic and biogeographic implications. *Cretaceous Research* **13**, 207–262.

Henderson RA, Crampton JS, Dettmann ME, Douglas JG, Haig D, Shafik S, Stilwell JD and Thulborn RA 2000. Biogeographical observations on the Cretaceous biota of Australasia. *Memoirs of the Australasian Association of Palaeontologists* **23**, 355–404.

McGloughlin S, Drinnan AN and Rozefelds AC 1995. A Cenomanian flora from the Winton Formation, Eromanga Basin, Queensland, Australia. *Memoirs of the Queensland Museum* **38**, 273–313.

Nagalingum NS, Drinnan AN, Lupia R and McLoughlin S 2002. Fern spore diversity and abundance in Australia during the Cretaceous. *Review of Palaeobotany and Palynology* **119**, 69–92.

Peters MD and Christophel DC 1978. *Austrosequoia wintonensis*, a new taxodiaceous cone from Queensland, Australia. *Canadian Journal of Botany* **56**, 3119–3128.

Pole MS 1999. Latest Albian-earliest Cenomanian monocotyledonous leaves from Australia. *Botanical Journal of the Linnean Society* **129**, 177–186.

Pole MS 2000. Dicotyledonous leaf macrofossils from the latest Albian-earliest Cenomanian of the Eromanga Basin, Queensland, Australia. *Paleontological Research* **4**, 39–52.

Pole MS 2000. Mid-Cretaceous conifers from the Eromanga Basin, Australia. *Australian Systematic Botany* **13**, 153–197.

Pole MS and Douglas JG 1999. Bennettitales, Cycadales and Ginkgoales from the mid Cretaceous of the Eromanga Basin, Queensland, Australia. *Cretaceous Research* **20**, 523–538.

Late Cretaceous marine and non-marine invertebrates

Cook AG 2005. First record of fossil freshwater gastropods within the Winton Formation. *Memoirs of the Queensland Museum* **51**, 406.

Darragh TA and Kendrick GW 1991. Maastrichtian Bivalvia (excluding Inoceramidae) from the Miria Formation, Carnarvon Basin, north-western Australia. *Records of the Western Australian Museum, Supplement* **36**, 1–102.

Darragh TA and Kendrick GW 1994. Maastrichtian Scaphopoda and Gastropoda from the Miria Formation, Carnarvon Basin, north-western Australia. *Records of the Western Australian Museum, Supplement* **48**, 1–76.

Etheridge R 1913. The Cretaceous fossils of the Gingin Chalk. *Geological Survey of Western Australia, Bulletin* **55**, 9–31.

Feldtmann FR 1951. Pectens of the Gingin Chalk 3. *Journal of the Royal Society of Western Australia* **5**, 9–39.

Feldtmann FR 1963. Some pelecypods from the Cretaceous Gingin Chalk, Western Australia, together with descriptions of the principal chalk exposures. *Journal of the Royal Society of Western Australia* **46**, 101–125.

Glenister BF, Miller AK and Furnish WM 1956. Upper Cretaceous and Early Tertiary nautiloids from Western Australia. *Journal of Paleontology* **30**, 492–503.

Henderson RA 1990. Late Albian ammonites from the Northern Territory, Australia. *Alcheringa* **14**, 109–148.

Henderson RA and McNamara KJ 1985. Taphonomy and ichnology of cephalopod shells in a Maastrichtian chalk from Western Australia. *Lethaia* **18**, 305–322.

Henderson RA and McNamara KJ 1985. Maastrichtian non-heteromorph ammonites from the Miria Formation, Western Australia. *Palaeontology* **28**, 35–88.

Henderson RA, Crampton JS, Dettmann ME, Douglas JG, Haig D, Shafik S, Stilwell JD and Thulborn RA 2000. Biogeographical observations on the Cretaceous biota of Australasia. *Memoirs of the Australasian Association of Palaeontologists* **23**, 355–404.

Henderson RA, Kennedy WJ and McNamara KJ 1992. Maastrichtian heteromorph ammonites from the Carnarvon Basin, Western Australia. *Alcheringa* **16**, 133–170.

Hocknull SA 1997. Cretaceous freshwater bivalves from Queensland. *Memoirs of the Queensland Museum* **42**, 223–226.

Hocknull SA 2000. Mesozoic freshwater and estuarine bivalves from Australia. *Memoirs of the Queensland Museum* **45**, 405–426.

Jell PA 2004. Fossil insects of Australia. *Memoirs of the Queensland Museum* **50**, 1–123.

McNamara KH, Friend D and Long J 1993. *A Guide to Fossils of the Gingin Chalk.* Western Australian Museum, Perth.

McNamara KH, Rexilius JP, Marshall NG and Henderson RA 1988. The first record of a Maastrichtian ammonite from the Perth Basin, Western Australia, and its biostratigraphical significance. *Alcheringa* **12**, 163–168.

Wright CW 1963. Cretaceous ammonites from Bathurst Island, northern Australia. *Palaeontology* **6**, 597–614.

Late Cretaceous sharks and fish

Cappetta H 1987. *Chondrichtyes II: Mesozoic and Cenozoic Elasmobranchii.* Handbook of Paleoichthyology, Vol. 3B. Gustav Fisher Verlag, Stuttgart.

Kemp N 1991. Chondrichthyans in the Cretaceous and Tertiary of Australia. In *Vertebrate Palaeontology of Australasia.* (Eds Vickers-Rich P, Monaghan JM, Baird RF and Rich TH) pp. 497–568. Pioneer Design Studio, Monash University, Melbourne.

Pledge NS 1992. Fossil shark teeth dredged from the Great Australian Bight. *Bureau of Mineral Reseources, Journal of Australian Geology and Geophysics* **13**, 15–18.

Siverson M 1996. Lamniform sharks of the Mid-Cretaceous Alinga Formation and Beedagong Claystone, Western Australia. *Palaeontology* **39**, 813–849.

Siverson M 1997. Sharks from the Mid-Cretaceous Gearle Siltstone, southern Carnarvon Basin, Western Australia. *Journal of Vertebrate Paleontology* **17**, 453–465.

Siverson M 1999. A new large lamniform shark from the uppermost Gearle Siltstone (Cenomanian Late Cretaceous) of Western Australia. *Transactions of the Royal Society of Edinburgh: Earth Sciences* **90**, 49–66.

Late Cretaceous non-marine turtles

Hocknull SA, White MA, Tischler TR, Cook AG, Calleja ND, Sloan T and Elliott DA 2009. New Mid-Cretaceous (Latest Albian) Dinosaurs from Winton, Queensland, Australia. *PLoS ONE 4*, e6190. doi:10.1371/journal.pone.0006190.

Molnar RE 1991. Fossil reptiles in Australia. In *Vertebrate Palaeontology of Australasia.* (Eds Vickers-Rich P, Monaghan JM, Baird RF and Rich TH) pp. 605–702. Pioneer Design Studio, Monash University, Melbourne.

Late Cretaceous ichthyosaurs

Choo B 1999. Cretaceous ichthyosaurs from Western Australia. *Records of the Western Australian Museum, Supplement* **57**, 207–218.

Teichert C and Matheson RS 1944. Upper Cretaceous ichthyosaurian and plesiosaurian remains from Western Australia. *Australian Journal of Science* **6**, 167–178.

Late Cretaceous plesiosaurs

Kear BP 2003. Cretaceous marine reptiles of Australia: a review of taxonomy and distribution. *Cretaceous Research* **24**, 277–303.

Long JA and Cruickshank ARI 1998. Further records of plesiosaurian reptiles of Jurassic and Cretaceous age from Western Australia. *Records of the Western Australian Museum* **19**, 47–55.

Late Cretaceous marine lizards (mosasaurs and dolichosaurs)

Kear BP, Long JA and Martin JE 2005. A review of Australian mosasaur occurrences. *Netherlands Journal of Geosciences* **84**, 307–313.

Lundelius E and Warne SStJ 1960. Mosasaur remains from the Upper Cretaceous of Western Australia. *Journal of Paleontology* **34**, 1215–1217.

Scanlon JD and Hocknull SA 2008. A dolichosaurid lizard from the latest Albian (mid-Cretaceous) Winton Formation, Queensland, Australia. *Transactions of the Kansas Academy of Science, Fort Hays Studies Special Issue* **3**, 131–136.

Late Cretaceous crocodiles

Salisbury SW, Molnar RE, Frey E and Willis PMA 2006. The origin of modern crocodyliforms: new evidence from the Cretaceous of Australia. *Proceedings of the Royal Society, Series B* **273**, 2439–2448.

Late Cretaceous dinosaurs

Coombs WP Jr and Molnar RE 1981. Sauropoda (Reptilia, Saurischia) from the Cretaceous of Queensland. *Memoirs of the Queensland Museum* **20**, 351–373.

Hocknull SA and Cook AG 2008. Hypsilophodontid (Dinosauria, Ornithischia) from latest Albian, Winton Formation, central Queensland. *Memoirs of the Queensland Museum* **52**, 212.

Hocknull SA, White MA, Tischler TR, Cook AG, Calleja ND, Sloan T and Elliott DA 2009. New Mid-Cretaceous (Latest Albian) Dinosaurs from Winton, Queensland, Australia. *PLoS ONE 4*, e6190. doi:10.1371/journal.pone.0006190.

Long JA 1992. First dinosaur bones from Western Australia. *The Beagle* **9**, 21–27.

Long JA 1992. New finds of Mesozoic reptiles from Western Australia. *The Beagle* **9**, 261–262.

Long JA 1995. A theropod bone from the Late Cretaceous Molecap Greensand, Western Australia. *Records of the Western Australian Museum* **17**, 143–146.

Molnar RE and Salisbury SW 2005. Observations on Cretaceous sauropods from Australia. In *Thunder-Lizards: The Sauropodomorph Dinoaurs.* (Eds Tidwell V and Carpenter K) pp. 454–465. Indiana University Press, Bloomington.

Thulborn RA and Wade M 1979. Dinosaur stampede in the Cretaceous of Queensland, Australia. *Lethaia* **12**, 275–279.

Thulborn RA and Wade M 1984. Dinosaur trackways in the Winton Formation (mid-Cretaceous) of Queensland (Australia). *Memoirs of the Queensland Museum* **21**, 413–518.

Late Cretaceous pterosaurs

Bennett SC and Long JA 1991. A large pterodactyloid pterosaur from the Late Cretaceous (Late Maastrichtian) of Western Australia. *Records of the Western Australian Museum* **15**, 435–443.

Hocknull SA and Cook AG 2007. Flights of fancy! On the trail of the Aussie pterosaurs. *Australian Age of Dinosaurs* **5**, 26–35.

INDEX

www.ingramcontent.com/pod-product-compliance
Ingram Content Group UK Ltd.
Pitfield, Milton Keynes, MK11 3LW, UK
UKHW060012051026
15591UKWH00002B/161

* 9 7 8 0 6 4 3 1 0 0 4 5 9 *